SMP interact

C3

Teacher's guide to Book C3

CAMBRIDGE
UNIVERSITY PRESS

PUBLISHED BY THE PRESS SYNDICATE OF THE UNIVERSITY OF CAMBRIDGE
The Pitt Building, Trumpington Street, Cambridge, United Kingdom

CAMBRIDGE UNIVERSITY PRESS
The Edinburgh Building, Cambridge CB2 2RU, UK
40 West 20th Street, New York, NY 10011–4211, USA
477 Williamstown Road, Port Melbourne, VIC 3207, Australia
Ruiz de Alarcón 13, 28014 Madrid, Spain
Dock House, The Waterfront, Cape Town 8001, South Africa

http://www.cambridge.org

Printed in the United Kingdon at the University Press, Cambridge

Typeface Minion   System QuarkXPress®

A catalogue record for this book is available from the British Library.

ISBN 0 521 78533 2   paperback

Typesetting and technical illustrations by The School Mathematics Project
Cover image © ImageState Ltd
Cover design by Angela Ashton

Contents

The following people contributed to the writing of the SMP Interact key stage 3 materials.

Ben Alldred	Ian Edney	John Ling	Susan Shilton
Juliette Baldwin	Steve Feller	Carole Martin	Caroline Starkey
Simon Baxter	Rose Flower	Peter Moody	Liz Stewart
Gill Beeney	John Gardiner	Lorna Mulhern	Pam Turner
Roger Beeney	Bob Hartman	Mary Pardoe	Biff Vernon
Roger Bentote	Spencer Instone	Peter Ransom	Jo Waddingham
Sue Briggs	Liz Jackson	Paul Scruton	Nigel Webb
David Cassell	Pamela Leon	Richard Sharpe	Heather West

Others, too numerous to mention individually, gave valuable advice, particularly by commenting on and trialling draft materials.

Editorial team:	David Cassell	Project Administrator:	Ann White
	Spencer Instone	Design:	Pamela Alford
	John Ling		Melanie Bull
	Paul Scruton		Nicky Lake
	Susan Shilton		Tiffany Passmore
	Caroline Starkey		Martin Smith
	Heather West	Project support:	Carol Cole
			Pam Keetch
			Jane Seaton
			Cathy Syred

Special thanks go to Colin Goldsmith.

Introduction

What is distinctive about *SMP Interact*?

SMP Interact sets out to help teachers use a variety of teaching approaches in order to stimulate pupils and foster their understanding and enjoyment of mathematics.

A central place is given to discussion and other interactive work. Through discussion with the whole class you can find out about pupils' prior understanding when beginning a topic, can check on their progress and can draw ideas together as work comes to an end. Working interactively on some topics in small groups gives pupils, including the less confident, a chance to clarify and justify their own ideas and to build on, or raise objections to, suggestions put forward by others.

Questions that promote effective discussion and activities well suited to group work occur throughout the material.

SMP Interact has benefited from extensive and successful trialling in a variety of schools. The practical suggestions contained in the teacher's guides are based on teachers' experiences, often expressed in their own words.

Who are the C Books for?

They are for use in key stage 3 and cover national curriculum levels up to 7. (Level 8 can be covered from the Higher tier GCSE book.)

How are the pupils' books intended to be used?

The pupils' books are a resource which can and should be used flexibly. They are not for pupils to work through individually at their own pace. Many of the activities are designed for class or group discussion.

Activities intended to be led by the teacher are shown by a solid strip at the edge of the pupil's page, and a corresponding strip in the margin of the teacher's guide, where they are fully described.

A broken strip at the edge of the page shows an activity or question in the pupil's book that is likely to need teacher intervention and support.

Where the writers have a particular way of working in mind, this is stated (for example, 'for two or more people').

Where there is no indication otherwise, the material is suitable for pupils working on their own.

Starred questions (for example, *C7) are more challenging.

What use is made of software?

Points at which software (on a computer or a graphic calculator) can be used to provide effective support for the work are indicated by these symbols, referring to a spreadsheet, graph plotter or dynamic geometry package respectively. Other suggestions for software support can be found on the SMP's website: www.smpmaths.org.uk

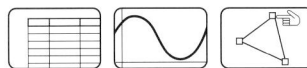

How is the attainment of pupils assessed?

The interactive class sessions provide much feedback to the teacher about pupils' levels of understanding.

Each unit of work begins with a statement of the key learning objectives and finishes with questions for self-assessment ('What progress have you made?'). The latter can be incorporated into a running record of progress.

Revision questions are included in periodic reviews in the pupil's book.

Assessment materials in the form of photocopiable masters with editable files on a CD-ROM are being developed. Details of the pack are on the SMP's website.

What will pupils do for homework?

The practice booklets may be used for homework.

Often a homework can consist of preparatory or follow-up work to an activity in the main pupil's book.

Answers to questions on resource sheets

For reasons of economy, where pupils have to write their responses on a resource sheet the answers are not always shown in this guide. For convenience in marking you could put the correct responses on a spare copy of each sheet and add it to a file for future use.

General guidance on teaching approaches

Getting everyone involved

When you are introducing a new idea or extending an already familiar topic, it is important to get as many pupils as possible actively engaged.

Posing key questions

A powerful technique for achieving this is to pose one or two key questions, perhaps in the form of a novel problem to be solved. Ask pupils, working in pairs or small groups, to think about the question and try to come up with an answer.

When everyone has had time to work seriously at the problem (have a further question ready for the faster ones), you can then ask for answers, without at this stage revealing whether they are right or wrong (so that pupils have to keep thinking!). You could ask pupils to comment on each other's answers.

Open tasks

Open tasks and questions are often good for getting pupils to think, or thinking more deeply. For example, 'Working in groups of three or four, make up three questions which can be solved using the technique we have just been learning. Try to make your questions as varied as possible.'

Questioning skills

Questioning with the whole class

If your questions to the class are always closed, and you reward the first correct response you get, then you have no way of telling whether other pupils knew the correct answer or whether they had thought about the question at all. It is better to try to get as many pupils as possible to engage with the question, so do not at first say whether an answer is right or wrong. You could ask a pupil how they got their answer, or you could ask a second pupil how they think the first one got their answer.

Working in groups

Types of group work

Group work may be small scale or large scale. In small scale group work, pupils are asked to work in pairs or small groups for a short while, perhaps to come up with a solution to a novel type of problem before their suggestions are compared. In large scale group work, pupils carry out in groups a substantial task such as planning a statistical inquiry or designing a poster to get over the essential idea of the topic they have just been studying.

Organising the groups

Group size is important. Groups of more than four or five can lead to some pupils making little or no contribution.

For some activities, you may want pupils to work unassisted. But for many, your own contribution will be vital. Then it is generally more effective if, once you are sure that every group has got started, you work intensively with each group in turn.

After the group work One way to help pupils feel that they have worked effectively is to get them to report their findings to the whole class. This may be done in a number of different ways. One pupil from each group could report back. Or you could question each group in turn. Or each group could make a poster showing their results.

Managing discussion

Discussion, whether in a whole-class or group setting, has a vital role to play in developing pupils' understanding. It is most fruitful in an atmosphere where pupils know their contributions are valued and are not always judged in terms of immediate correctness. It needs careful management for it to be effective and teachers are often worried that it will get out of hand. Here are a few common worries, and ways of dealing with them.

What if … '… the group is not used to discussion?'

- Allow time for pupils to work first on the problem individually or in small groups, then they will all have ideas to contribute.

'… everyone tries to talk at once?'

- Set clear rules. For example, pupils raise their hands and you write their name on the board before they can speak.

'… a few pupils dominate whole-class discussion?'

- Precede any class discussion with small-group discussion and nominate the pupils who will feed back to the class.

'… one pupil reaches the end point of a discussion immediately?'

- Tell them that the rest of the group need to be convinced and ask the pupil to convince the rest of the group.

- Accept the suggestion and ask the rest of the group to comment on it.

① Circumference of a circle

Essential

Cylindrical can and lipstick or correcting fluid

Practice booklet pages 3 to 6

A How many times? (p 4)

◊   The two diagrams show that the circumference is between 3 times and 4 times the diameter, but closer to 3 times.

◊   You could give a practical demonstration by wrapping a paper strip around a cylindrical can, unwrapping it and then stepping off the diameter of the can on the strip.

B Becoming more accurate (p 6)

C Calculating diameter and radius (p 9)

C2   This question highlights a common error and is worth discussing as a class.

D Rolling and turning (p 10)

Cylindrical can and lipstick or correcting fluid

◊   Pupils sometimes find it difficult to relate the circumference of a rolling object to the distance rolled each turn. The practical demonstration here should make the connection clear.

◊ The more demanding questions in this section provide an opportunity to stress the need for setting out the steps of working carefully.

🄴 Parts of circles (p 11)

◊ An informal approach is intended, rather than the use of a formula for the length of an arc.

🄐 How many times? (p 4)

Answers close to these are acceptable.

A1 (a) 45 cm   (b) 22.5 cm   (c) 69 cm

A2 5.7 cm or 6 cm

A3 (a) 28.5 cm   (b) 10.8 cm

A4 9000 cm (90 metres)

A5 12 cm

A6 45 cm

A7 380 m

A8 8000 miles

A9 1 metre

A10 90 m

A11 (a) 6 m

    (b) Multiply the previous answer by 2, giving 12 m

    (c) 18 m

🄑 Becoming more accurate (p 6)

B1 Answers may differ slightly from these.

Diameter	× ?	Circumference
4 cm	3.15	12.6 cm
6 cm	3.15	18.9 cm
8 cm	3.14	25.1 cm
10 cm	3.14	31.4 cm
12 cm	3.14	37.7 cm
14 cm	3.14	44.0 cm

B2 (a) 13.8 cm   (b) 10.7 cm   (c) 8.2 cm
    (d) 5.7 cm   (e) 2.8 cm

B3 (a) 2.6 cm   (b) 8.2 cm

B4 (a) 11.9 cm   (b) 8.8 cm   (c) 22.6 cm
    (d) 5.0 cm   (e) 37.1 cm

B5 (a) 20.1 cm   (b) 80.4 cm   (c) 33.0 cm
    (d) 145.8 cm   (e) 47.8 cm   (f) 95.5 cm

B6 The measured diameter, and hence the circumference, may differ from these values slightly.

    (a) Diameter = 2.7 cm
        Circumference = 8.5 cm

    (b) Diameter = 2.5 cm
        Circumference = 7.9 cm

    (c) Diameter = 0.9 cm
        Circumference = 2.8 cm

    (d) Diameter = 3.9 cm
        Circumference = 12.3 cm

B7 (a) 11.31 m or 1131 cm
    (b) 0.63 m or 63 cm

B8 50 m further
(inner circumference = 628.3 m, outer circumference = 678.6 m)

🄲 Calculating diameter and radius (p 9)

C1 (a) 1.6 cm   (b) 2.3 cm   (c) 2.7 cm
    (d) 3.1 cm

C2 The sequence is not correct.
It should be either

or

| 5 | 8 | ÷ | (| 2 | × | π |) | = |

| 5 | 8 | ÷ | 2 | ÷ | π | = |

C3 (a) 10.7 cm (b) 23.2 cm (c) 7.1 cm
(d) 16.9 cm (e) 1.5 cm

C4 6.4 cm

C5 (a) P (b) Q

C6 1.69 m

Ⓓ **Rolling and turning** (p 10)

D1 (a) 126 cm (b) 126 m
(c) 2.1 metres per second

D2 15.9 cm

D3 (a) 1.51 m (b) 663 turns

D4 65.9 cm

D5 11 turns

*D6 0.28 metre per second

Ⓔ **Parts of circles** (p 11)

E1 61 cm

E2 15.2 cm

E3 51.8 cm

E4 44.0 cm

*E5 4.8 cm

The pupil's informal explanation,
equivalent to the following

$$\frac{\pi d}{2} + d = 12.3$$

$$d\left(\frac{\pi}{2} + 1\right) = 12.3$$

$$d = \frac{12.3}{\frac{\pi}{2} + 1} = 4.7845...$$

*E6 53.4 cm

The pupil's informal explanation,
equivalent to the following

Total of circumferences
$$= \pi d_1 + \pi d_2 + \pi d_3 = \pi(d_1 + d_2 + d_3)$$
$$= 17\pi = 53.407...$$

What progress have you made? (p 12)

1 Diameter = 3 cm, radius = 1.5 cm

2 6283 m (or an answer close to this)

3 17.6 cm

4 11.8 cm

5 15.9 m

6 226 cm

7 4.4 metres per second

8 15.0 cm

Practice booklet

Section A (p 3)

1 (a) 15 cm (b) 36 cm (c) 51 cm

2 (a) 96 cm (b) 72 cm (c) 48 cm

3 66 m

4 90 kerbstones

5 (a) 13.2 m (b) 1320 m

6

Type of tree	Diameter	Radius
Oak	0.4 m	0.2 m
Silver birch	0.1 m	0.05 m
Horse chestnut	0.7 m	0.35 m
Yew	1.4 m	0.7 m
Beech	0.8 m	0.4 m

Section B (p 4)

1 (a) 37.7 cm (b) 26.4 cm (c) 31.4 cm
(d) 23.2 cm (e) 51.8 cm

2 23.6 cm

3 (a) 44.0 cm   (b) 47.1 cm   (c) 94.2 cm

   (d) 2.5 cm   (e) 235.6 cm

   (f) 565.5 cm

4 15.7 m

Section C (p 5)

1 (a) 7.2 cm

   (b) (i) 8.1 cm   (ii) 9.8 cm   (iii) 11.3 cm

2 (a) 2.9 cm   (b) 8.8 cm   (c) 1.2 cm

   (d) 40.6 cm

3 (a) 1754 km   (b) 14 km

Section D (p 6)

1 The circumference is about 42.7 cm, so it will run about 8540 cm or 85 metres.

2 (a) 214 cm

   (b) The circumference is 2.14… metres, so it will go round about $1000 \div 2.14 = 467$, or 470 times.

3 (a) Circumference = 2.2305… metres, so it will go round $1000 \div 2.2305…$ = 448, or 450 times.

   (b) Circumference = 6.597… metres, so it will go round $1000 \div 6.597…$ = 152, or 150 times.

   (c) Diameter = 9 mm = 0.009 m, so circumference = 0.02827… metres, so it will go round $1000 \div 0.02827…$ = 35 400, or 35 000 times.

4 Diameter of reel = 3.0 cm, so circumference = 9.4247… cm = 0.094247… metres. So there are about $100 \div 0.094247…$ = 1060, or 1100 turns.

Section E (p 6)

1 (a) Shape has 2 straight sides plus a circle of diameter 2.7 cm so perimeter = $2 \times 2.7 + \pi \times 2.7 =$ 13.9 or 14 cm.

   (b) Shape consists of 3 semicircles, of diameters 18, 12 and 18 + 12 cm = 30 cm. So perimeter = 94 cm.

   (c) Shape is $\frac{1}{6}$ of a circle, radius 10 cm, plus 2 sides of 10 cm. So perimeter = 30.5 cm.

*2 If the radius is r cm, then the circumference of the three-quarter circle = $\frac{3}{4} \times 2 \times \pi \times r$ cm.
We also have 2 pieces of length r cm.
So $\frac{3}{4} \times 2 \times \pi \times r + 2 \times r = 20$
So $4.712… \times r + 2 \times r = 20$
So $6.712… \times r = 20$, $r = 2.979…$
The radius of the circle is 2.98 cm.

② Solving equations

> **Practice booklet** pages 7 to 9

𝔸 **Balancing** (p 13)

◊ In your introduction, emphasise that there are often many ways to solve equations. You could start with an equation and see what happens when you apply different operations on each side.

For example, start with $2x + 4 = 6$ and ask pupils about the result of:

- adding 2 to each side
- subtracting 2 from each side
- subtracting 4 from each side
- dividing both sides by 2

Ask which of these are more helpful in trying to solve the equation.

◊ On page 13 are two pupils' solutions for two equations. Without reference to this page, you could present pupils with these two equations and see what methods they come up with.

In solving $3x + 2 = x - 6$, Julie starts by adding 6. Pupils can discuss why they think she does this and whether it is the best way to start. Point out that even if you do not choose the most efficient 'first move' when solving an equation, you can still find a way to solve the equation.

Donal treats the right-hand side as $x + {}^-6$ and it may be helpful to be explicit about this at this stage.

◊ It is good practice for pupils to check their solutions. However, it can become very tedious, leading to a loss in momentum and motivation. You need to decide how far to encourage checking with your pupils.

A5 Pupils should use the method that seems most appropriate each time. If the division leads to a decimal or a fraction pupils may prefer to multiply out the brackets first. Otherwise, they may prefer to divide first.

A6/7 Pupils may prefer to use flowcharts to solve these equations. Parts (c) and (d) should help them appreciate that using reverse flowcharts is a method that has limitations.

B Solving problems (p 15)

◊ In B3 to B6, encourage pupils to solve each problem by forming an equation and solving it. They should avoid using 'trial and improvement'.

C Over-balancing? (p 16)

◊ You could begin your introduction by asking pupils to find some different ways to solve the first equation, $19 - 3x = 13$. Subtracting 13 and then adding $3x$ is very similar to the method shown on page 16. Another method is to start by subtracting 19 and then dividing by $^-3$. As before, encourage pupils to think about the most helpful 'first move' each time. In the second equation, $2(x - 3) = 9 - x$, discuss why multiplying out brackets might be preferred to dividing by 2 as a 'first move'.

D More problems (p 17)

D6 Henry Ernest Dudeney lived in a time when puzzles of all kinds were extremely popular. He was a self-taught mathematician and composed a huge variety of puzzles.
Nicolas Chuquet was a doctor and also the best French mathematician of his time.

E Other types of equation (p 19)

A Balancing (p 13)

A1 (a) $x = 5$ (b) $x = 3$
(c) $x = 1.5$ (d) $x = ^-4$

A2 He divided both sides by 3 and then added 1.

A3 (a) (i) $\quad 4(x + 3) = 28$
$\qquad\qquad 4x + 12 = 28$
$\qquad\qquad\qquad 4x = 16$
$\qquad\qquad\qquad\quad x = 4$

      (ii) $\quad 4(x + 3) = 28$
$\qquad\qquad\quad x + 3 = 7$
$\qquad\qquad\qquad x = 4$

(b) The pupil's preference. Donal's is quicker.

A4 (a) (i) $\quad 6(x - \frac{1}{2}) = 3$
$\qquad\qquad 6x - 3 = 3$
$\qquad\qquad\quad 6x = 6$
$\qquad\qquad\qquad x = 1$

      (ii) $\quad 6(x - \frac{1}{2}) = 3$
$\qquad\qquad\quad x - \frac{1}{2} = \frac{1}{2}$
$\qquad\qquad\qquad x = 1$

(b) The pupil's preference

A5 (a) $x = 16$ (b) $w = 1.5$ (c) $y = ^-3$
(d) $z = 5$ (e) $n = 1.2$ (f) $m = ^-2$

A6 $\dfrac{n+3}{4} = 2.5$

$n + 3 = 10$

$n = 7$

A7 (a) $p = 16$ (b) $n = {}^-3$

(c) $q = 4.5$ (d) $m = 10$

A8 (a) $x = {}^-10$ (b) $y = 2$ (c) $z = 14$

(d) $m = 5.5$ (e) $n = 1$ (f) $p = 2.5$

(g) $q = 28$ (h) $r = {}^-0.5$

B Solving problems (p 15)

B1 (a) 1 hour 20 minutes

(b) (i) The pupil's explanation

(ii) 4 kg

(c) $40w + 20 = 160$
3.5 kg

B2 (a) $f = 59$ (b) $c = 10$

(c) $c = 37.8$ (to 1 d.p.)

B3 16

B4 (a) 100 cm^2 (b) $l = 3$

*****B5** £4.50

*****B6** 4 years old

C Over-balancing? (p 16)

C1 (a) $x = 2$ (b) $x = {}^-5$

(c) $x = 6$ (d) $x = 3$

C2 (a) $k = 3$ (b) $j = 1$ (c) $h = 3.5$

(d) $g = {}^-1$ (e) $f = 1.5$ (f) $e = 1.8$

(g) $d = 6$ (h) $c = {}^-7$ (i) $b = {}^-1$

(j) $a = {}^-1.5$ (k) $c = 1$ (l) $d = 0$

C3 (a) $x = 1.5$ (b) $x = 5$ (c) $x = {}^-7$

(d) $x = 81$ (e) $x = {}^-1.8$ (f) $x = 16$

(g) $x = 5$ (h) $x = 3$ (i) $x = {}^-4$

C4 (a) $x = 9$ (b) $x = 3$ (c) $x = 2.5$

(d) $x = 3$ (e) $x = 6$ (f) $x = {}^-5$

D More problems (p 17)

D1 (a) Puzzle 1: C $(12 - 2n = 4n)$
Puzzle 2: B $(2n - 12 = 4n)$

(b) Puzzle 1: the number is 2
Puzzle 2: the number is $^-6$

In each solution, a different letter can be used for the unknown.

D2 (a) $21 - 3n = 4n$
$n = 3$

(b) $5 - 4n = n + 15$
$n = {}^-2$

(c) $3(2 - n) = n + 2$
$n = 1$

(d) $7(3 - 2n) = n$
$n = 1.4$

D3 (a) 24 (solving $12 - 2n = 3(12 - n)$)

(b) $^-36$

D4 (a) £3.50 (solving $16 - 3c = 9 - c$)

(b) £5.50

*****D5** (a) $5 - 2x$ miles

(b) 1:30 p.m. (solving $5 - 2x = 8 - 4x$)

(c) 2 miles from the crossroads

*****D6** (a) In four and a half years
(solving $45 + t = 3(12 + t)$)

(b) 18 days (solving $2w = 3(30 - w)$)

(c) 9:36 p.m.
(solving $\frac{1}{4}t + \frac{1}{2}(24 - t) = t$)

E Other types of equation (p 19)

E1 (a) $x = 6$ (b) $n = 2$ (c) $p = 2$

(d) $n = 3$ (e) $b = 16$ (f) $n = 12$

(g) $y = 3.5$ (h) $c = 0.5$ (i) $m = 0.1$

E2 (a) $n = 5$ or $^-5$ (b) $a = 8$ or $^-8$

(c) $f = 6$ or $^-6$ (d) $p = 3$ or $^-3$

(e) $n = 9$ or $^-9$ (f) $n = 4$ or $^-4$

(g) $x = 3$ or $^-3$ (h) $x = 2$

(i) $p = 3$ (j) $q = 0.5$ or $^-0.5$

(k) $n = 0.1$ (l) $x = 5$ or $^-5$

*E3 (a) 5 or $^-5$ (b) 6 or $^-6$ (c) 4 or $^-4$
(d) 3 or $^-3$ (e) 3   (f) 2 or $^-2$
(g) 6 or $^-7$

What progress have you made? (p 20)

1 (a) $v = 6$ (b) $w = 2.5$ (c) $x = ^-1$
(d) $y = 12$ (e) $z = 2$

2 (a) $p = 7$ (b) $q = 4$

3 (a) $m = 1.5$ (b) $n = ^-2$
(c) $g = ^-2$ (d) $h = 3$

4 $^-6$ (solving $5n + 2 = 2(n - 8)$)

5 4 (solving $36 - 6n = 3n$)

6 $^-10$ (solving $30 - 5n = 4(10 - n)$)

7 (a) $x = 9$ (b) $n = 4$ or $^-4$
(c) $n = 6$ or $^-6$ (d) $n = 0.25$

Practice booklet

Section A (p 7)

1 (a) $x = 6$ (b) $x = 1$
(c) $x = 4.4$ (d) $x = ^-7$

2 (a) $4x - 2 = 10$
$4x = 12$
$x = 3$
(b) $x - \frac{1}{2} = 2\frac{1}{2}$
$x = 3$

3 (a) $x = 17$ (b) $y = ^-4.5$
(c) $d = 1.4$ (d) $p = 3$

4 (a) $m = 41$ (b) $p = ^-4.5$
(c) $r = 2.4$ (d) $s = ^-25$

5 (a) $x = ^-5$ (b) $x = ^-2$
(c) $r = 1.8$ (d) $p = 12$

Section B (p 7)

1 (a) $c = 39$ (b) $b = 16\frac{1}{2}$
(c) $b = 12\frac{1}{2}$

2 $5x - 6 = 3(x + 12)$
$x = 21$
The answer was 99.

3 (a) Area of rectangle $= 6.3 \text{ cm}^2$
(Solve $3l + 7 = 28$ to find $l = 7$)
(b) $l = 3$ ($3l + 7 = 4l + 4$)

4 (Solve $2x - 13 = 3(x - 13)$
to find $x = 26$
where x is the age of the son.)
Mr Scott is 52.

Section C (p 8)

1 (a) $x = ^-1$ (b) $y = 2$
(c) $c = 1\frac{1}{2}$ (d) $f = ^-2$

2 (a) $x = ^-0.25$ (b) $y = 2$
(c) $b = ^-0.75$ (d) $f = 2.5$
(e) $p = 0.8$ (f) $h = ^-3$

3 (a) $x = 2.75$ (b) $x = ^-0.75$
(c) $x = ^-3$ (d) $x = 36$
(e) $x = 28$ (f) $x = ^-0.5$
(g) $x = 6$ (h) $x = ^-13$

Section D (p 9)

1 8 (solving $16 - 3n = 5n$)

2 (a) 14 (solving $3s + 2 = s + 30$)
(b) 44

3 5 (solving $3(20 - n) = 9n$)

4 $x = 3$, $y = 5$
Area $= 70 \text{ cm}^2$, perimeter $= 38 \text{ cm}$

③ Rates

> **Practice booklet** pages 10 to 12

Ⓐ **Rates of flow** (p 21)

◊ You could first ask what units would be used to measure a rate of flow. You could then ask pupils, working in pairs, to find the rate of flow of each tap whose graph is given.

Discussion of their methods should lead to the fact that the gradient, measured using the graph scales, gives the rate of flow.

◊ The rates of flow in litres per minute are

      A 30   B 12   C 7   D 4.4   E 1.25

Ⓑ **Travel graphs** (p 24)

◊ Before tackling the gradients, ask questions to make sure that the graph is well understood. For example:
  • How far was Donal from home at 10:30? at 13:30?
  • How far was he from the seaside at each of these times?
  • How long did he spend at the seaside?

Then you could ask whether he was faster going there or coming back, and how we can tell this from the graph.

◊ The speeds shown in the graph are 48 m.p.h. and 40 m.p.h.

Ⓒ **Problems involving rates** (p 26)

◊ The worked example illustrates the technique of replacing 'hard' by 'easy' numbers to see what operation is involved. More able pupils may appreciate a more algebraic approach. If R = rate, Q = quantity and T = time, then the definition of rate gives $R = Q/T$, from which may be derived $Q = RT$ and $T = Q/R$.

A Rates of flow (p 21)

A1 (a) 28 litres per minute
(b) 14 litres per minute
(c) 8 litres per minute
(d) 3.6 litres per minute

A2 (a) 5 litres per minute
(b) 17 minutes

A3 (a) Indoor 6 litres per minute
Outdoor 6.67 litres per minute
(b) Indoor 8.33 minutes
Outdoor 7.5 minutes

A4 11.1 minutes

A5 (a) Water is running out of a container at 2.5 litres per minute.
(b) 8 litres per minute

A6 (a) 8 litres per minute
(b) 18 litres per minute
(c) 10 litres per minute
(d) 5 litres per minute
(e) 21 minutes

B Travel graphs (p 24)

B1 (a) 15 miles
(b) A 30 m.p.h. B 20 m.p.h.

B2 (a) 6.7 m.p.h.   (b) 14.3 m.p.h.

B3 (a) She stopped at 7:30 for 30 minutes.
(b) 40 m.p.h.   (c) 30 m.p.h.
(d) 1 hour   (e) 33.3 m.p.h.

B4 (a) 32 m.p.h.   (b) 45 minutes
(c) 16 m.p.h.

B5 (a)

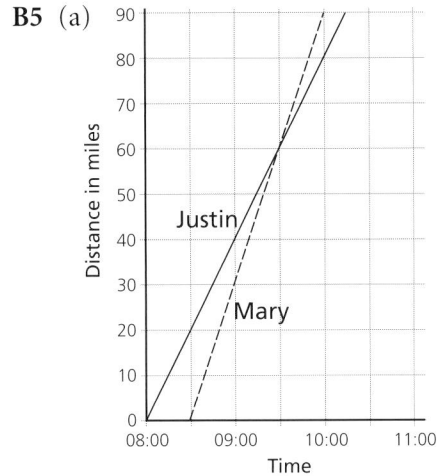

(b) 60 m.p.h.
(c) 09:30, 60 miles from home

***B6** 12:10

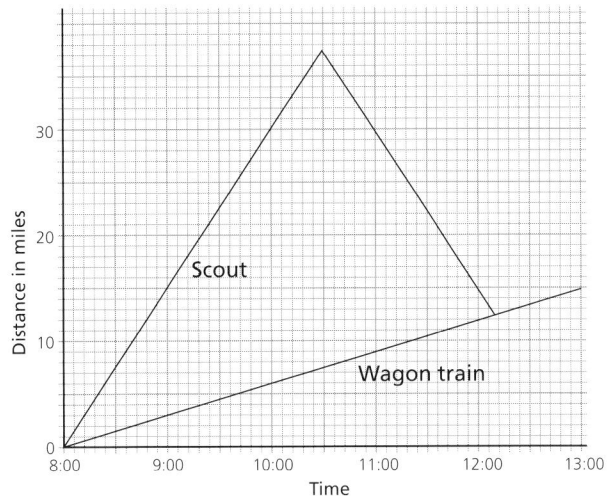

C Problems involving rates (p 26)

C1 (a) 10 litres   (b) 25 seconds

C2 (a) 144 litres (b) 0.63 litres per second
(c) 173.75 seconds

C3 (a) 4.8 litres per second
(b) 96 litres   (c) 94 seconds

C4 48 days (47.4 days)

C5 13.3 minutes (Karl's is faster)

C6 About 79 or 80 minutes

C7 (a) 6280 persons per km^2

(b) 5040 persons per km^2

*C8 9.76 minutes

What progress have you made (p 27)

1 (a) 7.5 litres per minute

(b) 2.5 litres per minute

2

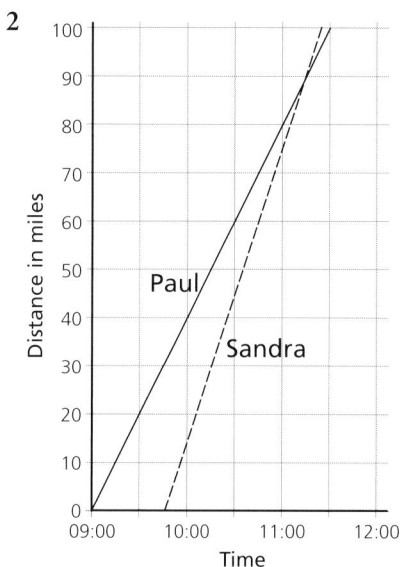

Sandra overtakes Paul at 11:15, 90 miles from where they joined the motorway.

3 4 minutes

Practice booklet

Section A (p 10)

1 (a) 8.6 litres per minute

(b) 7.5 litres per minute

2 (a) 18 litres per minute

(b) $3\frac{1}{3}$ minutes (3 minutes 20 seconds)

3 (a) 80 people per minute

(b) 3 exits

4 3000 sweets produced in
$2\frac{1}{2}$ hours = 1200 per hour
$1200 \div 450 = 2.7$ so 3 wrapping machines are needed.

Section B (p 11)

1 (a) 09:30   (b) 60 m.p.h.

(c) 26.7 m.p.h.   (d) 3 hours

(e) 46.7 m.p.h.

2

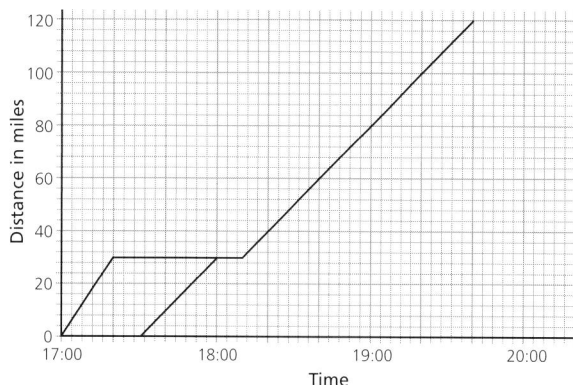

The train arrives in Birmingham at 19:40.

Section C (p 12)

1 51 litres, 2.43 litres per hour, 5.36 hours

2 About $10\frac{1}{2}$ minutes less
(Kay types faster)

3 31 m^2 should need $31 \div 14 = 2.21\ldots$ litres. So he used more. (Or: 2.4 litres should cover $2.4 \times 14 = 33.6$ m^2. So he used more.)

4 £7.11 per square yard

5 (a) 15.0 metres per minute

(b) 10.5 metres per minute

(c) 1 hour 18 minutes

④ Vectors

A vector is a quantity that has magnitude and direction. One example of a vector is a movement (or slide) without turning, which can be represented by two numbers in 'column vector' notation, and this is what is introduced here, leading to simple problems designed to develop flexible thinking.

Essential

Sheet 246 (one per pair of pupils)
Cards made from sheet 247 (one per pair)
Counters
Squared paper

Practice booklet pages 13 to 15

Ⓐ **Writing and drawing vectors** (p 28)

Sheet 246 (one per pair of pupils)
Cards made from sheet 247 (one per pair)
Counters
Squared paper

◊ The vectors **a** to **r** are intended to be used for teacher-led oral questions.

◊ 'Vector snakes and ladders' provides self-checking practice in relating a column vector to the 'move' it represents. It also serves to strengthen the idea that a vector can operate anywhere on the grid: it does not have fixed start and finish points.

The basic game is one of pure chance, but the strategy game allows tactics to be developed.

> **T**
>
> *'This was excellent. My group got very animated and learnt the four directions with +/– quicker than any other group that I have taught.'*

Ⓑ **Combining vectors** (p 30)

Squared paper

◊ Vector addition emerges from the idea of a journey consisting of several vectors. Pupils soon realise they do additions of the top and bottom numbers separately; the work gives practice in (and a supportive way of thinking about) addition of directed numbers.

◊ Questions B3 to B6 can be tackled by pupils working in groups of two to four. You can stop the groups from time to time and ask group representatives to explain their reasoning in choosing the vectors.

Ⓐ Writing and drawing vectors (p 28)

A1

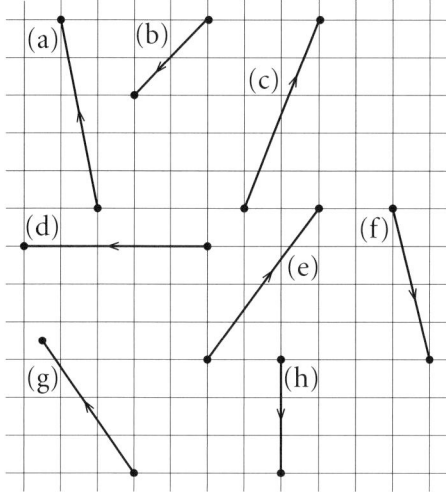

A2 $\mathbf{a} = \begin{bmatrix} 3 \\ 1\frac{1}{2} \end{bmatrix}$ $\mathbf{b} = \begin{bmatrix} 3 \\ -3 \end{bmatrix}$ $\mathbf{c} = \begin{bmatrix} -2 \\ -3 \end{bmatrix}$

$\mathbf{d} = \begin{bmatrix} 2 \\ -3\frac{1}{2} \end{bmatrix}$ $\mathbf{e} = \begin{bmatrix} -2\frac{1}{2} \\ -2\frac{1}{2} \end{bmatrix}$ $\mathbf{f} = \begin{bmatrix} -\frac{1}{2} \\ 4 \end{bmatrix}$

Ⓑ Combining vectors (p 30)

B1

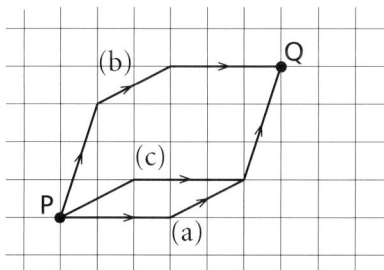

You always finish at Q, whatever order you draw the three vectors.

B2

The third vector is $\begin{bmatrix} 2 \\ -1 \end{bmatrix}$.

B3 (a) $\begin{bmatrix} 4 \\ 2 \end{bmatrix}$, $\begin{bmatrix} 2 \\ 3 \end{bmatrix}$ and $\begin{bmatrix} 3 \\ -1 \end{bmatrix}$

    (b) No, you can use them in any order.

B4 $\begin{bmatrix} 4 \\ 2 \end{bmatrix}$, $\begin{bmatrix} -1 \\ 2 \end{bmatrix}$ and $\begin{bmatrix} 7 \\ -1 \end{bmatrix}$

B5 (a) $\begin{bmatrix} 1 \\ -2 \end{bmatrix}$ and $\begin{bmatrix} 4 \\ 1 \end{bmatrix}$

    (b) $\begin{bmatrix} -1 \\ -2 \end{bmatrix}$, $\begin{bmatrix} -1 \\ 1 \end{bmatrix}$ and $\begin{bmatrix} 7 \\ 0 \end{bmatrix}$

B6 (a) $\begin{bmatrix} 5 \\ 9 \end{bmatrix}$ (b) $\begin{bmatrix} 2 \\ 5 \end{bmatrix}$ (c) $\begin{bmatrix} 5 \\ 3 \end{bmatrix}$

    (d) $\begin{bmatrix} 4 \\ -1 \end{bmatrix}$ (e) $\begin{bmatrix} -2 \\ 10 \end{bmatrix}$ (f) $\begin{bmatrix} 8 \\ -9 \end{bmatrix}$

B7 (a) $\begin{bmatrix} 11 \\ 3 \end{bmatrix}$ (b) $\begin{bmatrix} 3 \\ 2 \end{bmatrix}$ (c) $\begin{bmatrix} 11 \\ -3 \end{bmatrix}$

    (d) $\begin{bmatrix} -1 \\ 11 \end{bmatrix}$ (e) $\begin{bmatrix} -4 \\ 3 \end{bmatrix}$ (f) $\begin{bmatrix} 17 \\ 9 \end{bmatrix}$

    (g) $\begin{bmatrix} 6 \\ -6 \end{bmatrix}$ (h) $\begin{bmatrix} 17 \\ 3 \end{bmatrix}$ (i) $\begin{bmatrix} -4 \\ 9 \end{bmatrix}$

B8 $a = 1$ $b = 5$ $c = 6$
     $d = 0$ $e = 1$ $f = 2$
     $g = -4$ $h = -4$ $i = 3$
     $j = -8$ $k = -6$ $l = -1$

B9 (a) $\begin{bmatrix} -4 \\ 1 \end{bmatrix}$ (b) $\begin{bmatrix} -3 \\ 3 \end{bmatrix}$ (c) $\begin{bmatrix} -5 \\ -1 \end{bmatrix}$

    (d) $\begin{bmatrix} 3 \\ -10 \end{bmatrix}$ (e) $\begin{bmatrix} -12 \\ 2 \end{bmatrix}$ (f) $\begin{bmatrix} 14 \\ -5 \end{bmatrix}$

B10 (a) $\mathbf{j} = \begin{bmatrix} 3 \\ 4 \end{bmatrix}$ (b) $\mathbf{k} = \begin{bmatrix} -1 \\ 2 \end{bmatrix}$ (c) $\mathbf{l} = \begin{bmatrix} -5 \\ -11 \end{bmatrix}$

    (d) $\mathbf{m} = \begin{bmatrix} 2 \\ 1 \end{bmatrix}$ (e) $\mathbf{n} = \begin{bmatrix} -2 \\ 3 \end{bmatrix}$ (f) $\mathbf{o} = \begin{bmatrix} -3 \\ 3 \end{bmatrix}$

Round trip challenge

1

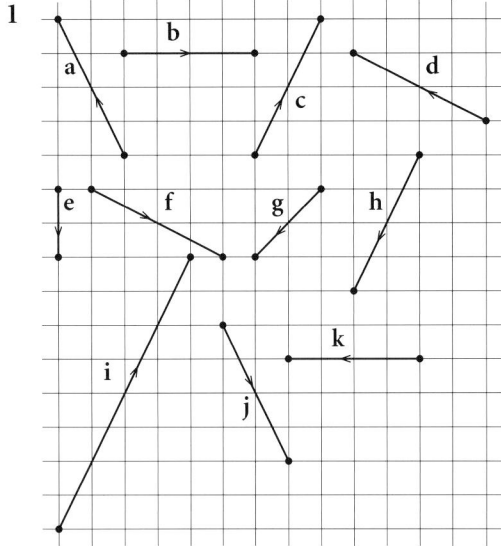

2 The top numbers add up to 0 and bottom numbers add up to 100.

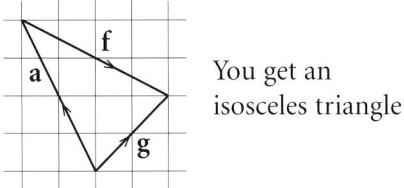

You get an isosceles triangle

3

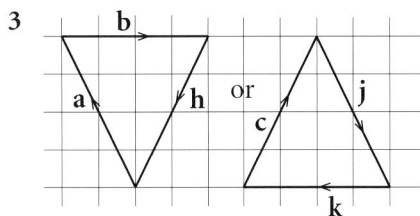

All the round trip journeys can be drawn the other way round. For example, the two above can be drawn like this.

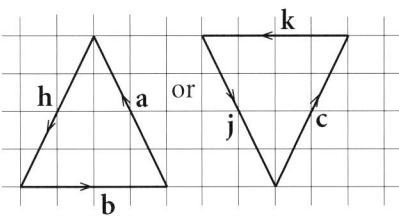

4

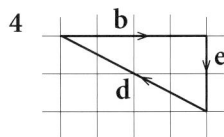

5 You need the vector $\begin{bmatrix} -2 \\ 0 \end{bmatrix}$.

This makes a kite.

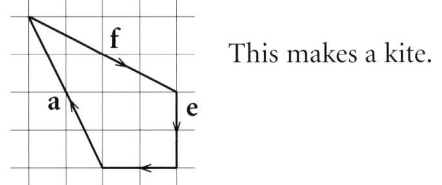

6

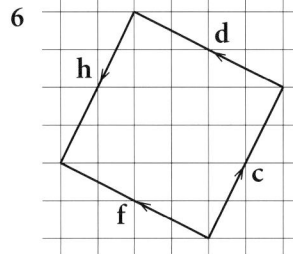

7 (a)

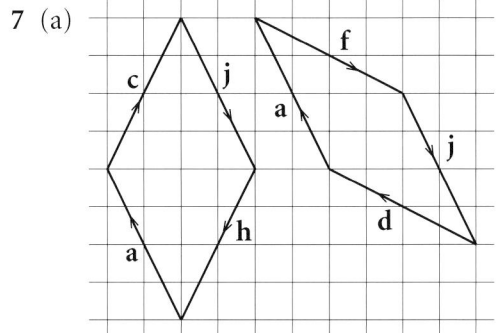

(b)

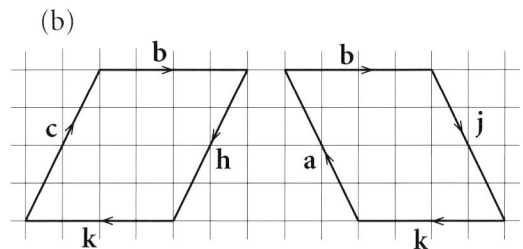

(c)

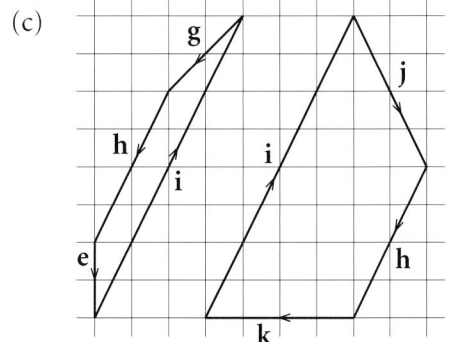

What progress have you made? (p 33)

1

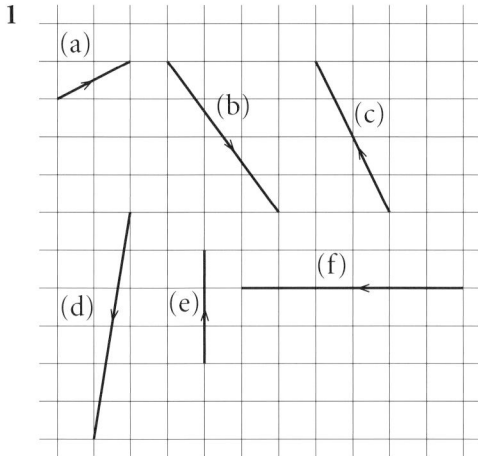

2 (a) $\begin{bmatrix} 2 \\ -3 \end{bmatrix}$ (b) $\begin{bmatrix} 5 \\ 2 \end{bmatrix}$ (c) $\begin{bmatrix} -3 \\ -3 \end{bmatrix}$

3 The pupil's choice of two vectors, for example
$$\begin{bmatrix} 5 \\ 0 \end{bmatrix} \text{ and } \begin{bmatrix} 3 \\ 2 \end{bmatrix}$$

4 $\begin{bmatrix} 2 \\ 6 \end{bmatrix}$

5 (a) $\begin{bmatrix} 5 \\ 2 \end{bmatrix}$ (b) $\begin{bmatrix} 1 \\ 8 \end{bmatrix}$

Practice booklet

Section A (p 13)

1 The pupil's drawings

2 (a) $\begin{bmatrix} -9 \\ 4 \end{bmatrix}$ (b) $\begin{bmatrix} -3 \\ -15 \end{bmatrix}$ (c) $\begin{bmatrix} -8 \\ 7 \end{bmatrix}$

     (d) $\begin{bmatrix} 4 \\ -10 \end{bmatrix}$ (e) $\begin{bmatrix} 0 \\ 9 \end{bmatrix}$ (f) $\begin{bmatrix} 2 \\ 0 \end{bmatrix}$

     (g) $\begin{bmatrix} -5 \\ -5 \end{bmatrix}$

3 $\mathbf{a} = \begin{bmatrix} 12 \\ 0 \end{bmatrix}$ $\mathbf{b} = \begin{bmatrix} 14 \\ 4 \end{bmatrix}$ $\mathbf{c} = \begin{bmatrix} -9 \\ 0 \end{bmatrix}$

    $\mathbf{d} = \begin{bmatrix} 10 \\ -7 \end{bmatrix}$ $\mathbf{e} = \begin{bmatrix} 0 \\ -12 \end{bmatrix}$ $\mathbf{f} = \begin{bmatrix} -14 \\ 4 \end{bmatrix}$

    $\mathbf{g} = \begin{bmatrix} 0 \\ 12 \end{bmatrix}$ $\mathbf{h} = \begin{bmatrix} -6 \\ -9 \end{bmatrix}$ $\mathbf{i} = \begin{bmatrix} -21 \\ 6 \end{bmatrix}$

    $\mathbf{j} = \begin{bmatrix} 15 \\ 15 \end{bmatrix}$

Section B (p 14)

1 (a) The vectors for the next three tacks
are $\begin{bmatrix} 3 \\ -6 \end{bmatrix}$, $\begin{bmatrix} 6 \\ 2 \end{bmatrix}$, $\begin{bmatrix} 4 \\ -8 \end{bmatrix}$.

    (b) Several correct pairs of vectors are possible for the last two tacks as long as the boat does not run ashore.

    (c) $\begin{bmatrix} 10 \\ -3 \end{bmatrix}$ (d) $\begin{bmatrix} 29 \\ -15 \end{bmatrix}$ (e) $\begin{bmatrix} -29 \\ 15 \end{bmatrix}$

2 (a) $\begin{bmatrix} 37 \\ 8 \end{bmatrix}$ (b) $\begin{bmatrix} -7 \\ 22 \end{bmatrix}$ (c) $\begin{bmatrix} -15 \\ -21 \end{bmatrix}$

    (d) $\begin{bmatrix} 0 \\ 0 \end{bmatrix}$

3 (a) 4   (b) -2   (c) 0   (d) 4, 7

4 $\mathbf{j} = \begin{bmatrix} 1 \\ 3 \end{bmatrix}$ $\mathbf{k} = \begin{bmatrix} 1 \\ -3 \end{bmatrix}$ $\mathbf{l} = \begin{bmatrix} -5 \\ -1 \end{bmatrix}$

   $\mathbf{m} = \begin{bmatrix} 17 \\ 4 \end{bmatrix}$ $\mathbf{n} = \begin{bmatrix} 13 \\ 8 \end{bmatrix}$ $\mathbf{o} = \begin{bmatrix} -6 \\ -4 \end{bmatrix}$

5 (a)

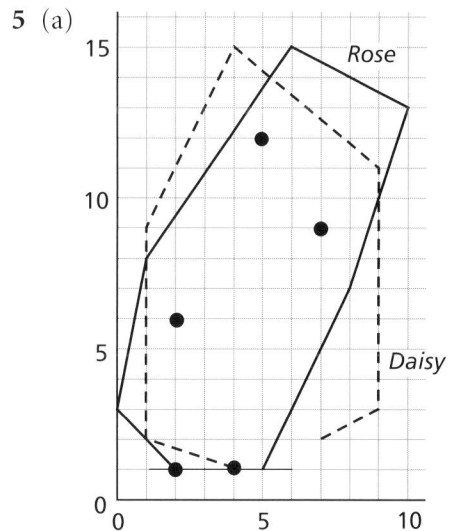

    (b) *Rose*

    (c) The final positions were *Rose* (5, 1) and *Daisy* (7, 2).

⑤ Manipulation

You may wish to cover the material in this unit a little at a time.

Essential	**Optional**
Sheets 248 and 249	Sheet 250
Practice booklet pages 16 to 21	

Ⓐ **Multiplying** (p 34)

> Sheet 248 for 'Cover up'

◊ The area diagrams give pupils a picture for multiplying two linear expressions. You could begin by presenting pupils with the three rectangles below and ask them to find expressions for the area of each.

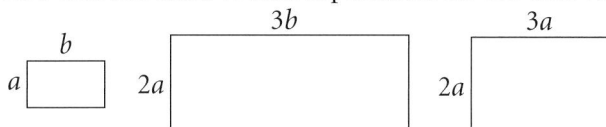

Then the illustrations on page 34 can be discussed and pupils could try A1.
Move on to more complex examples, such as $5d^2 \times 3d$ and $2a \times 5ab$, that are followed up in A2 and A3.

◊ Point out that, in products with more than two letters, the letters are often written in alphabetical order.

◊ You may wish to include work on negative numbers in your discussion, for example $^-2x \times 5x$, but it is not included in this unit.

◊ 'Cover up' is a puzzle that provides practice and can be set for homework.

B Dividing (p 36)

Sheet 249

◊ Initially pupils consider division as the inverse of multiplication. For example, solving the problem $3 \times ? = 12a$ is equivalent to finding $\frac{12a}{3}$.

When pupils reach B4, you may have to remind them that for example $\frac{15n}{5}$ can be solved by finding **?** in $5 \times ? = 15n$.

They go on to use the rules for simplifying fractions.

C Brackets (p 38)

◊ The diagrams can help to show the equivalence of expressions by considering areas.

Pupils may find area is not quite so illuminating a context in the case of subtraction. They may find it easier to consider the equivalence of, say, $2 \times 6 \times (100 - 4)$ and $2 \times 6 \times 100 - 2 \times 6 \times 4$ and generalise from there.

C1 Pupils can check their results for parts (a), (b), (c) and (g) by splitting up the rectangles and adding their areas. For example, the rectangle in part (a) can be split up as follows:

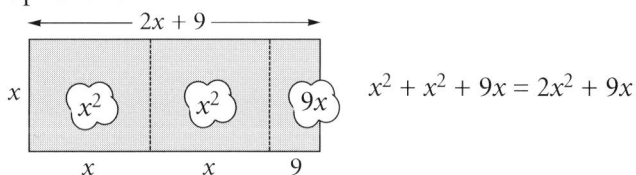

$$x^2 + x^2 + 9x = 2x^2 + 9x$$

The rectangle in part (d) can be considered in the same way as the second example at the top of the page but this may be less helpful.

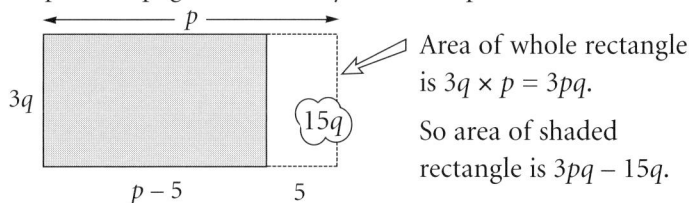

Area of whole rectangle is $3q \times p = 3pq$.

So area of shaded rectangle is $3pq - 15q$.

◊ Questions C4 and C5 provide a light introduction to factorising.

D True to form (p 40)

Pupils use the algebraic manipulation developed in earlier work to prove statements about multiples, odd, even, square and negative numbers.

◊ Pupils could work in groups to start with and report their findings on statements A to L to the class. They should try to justify their decisions.

At the end of the discussion, pupils should be able to use substitution and algebraic arguments to classify the statements on the page as always, sometimes or never true.

They should all understand the following types of argument:

- '$n + 10$ is negative' is **never** true if n is a positive integer. Two positive numbers added together will always produce a positive number.
- 'n^2 is a multiple of 5' is **sometimes** true. For example, if $n = 3$, $n^2 = 9$ which is not a multiple of 5. However, if $n = 5$, $n^2 = 25$ which is a multiple of 5.
- '$6n + 8$ is even' is **always** true. $6n + 8 = 2(3n + 4)$ and $3n + 4$ is an integer (because n is) so $6n + 8$ is always even.
- '$2n + 5$ is odd' is **always** true. $2n$ is always even and 5 is odd. An even added to an odd number will always make an odd number.

E Grid totals (p 41)

Pupils have the opportunity to use the work on multiples from section D in the context of investigating totals on a grid.

> Optional: sheet 250

E2 Pupils can investigate the effect of placing n in different squares.

F Ways of seeing (p 42)

◊ Pupils could work in groups to find different ways to count the red Smarties needed for design 10.

Ask them to think about how to use each way of counting the red Smarties to find the rule for the number of reds in the nth design. Some ways are shown below.

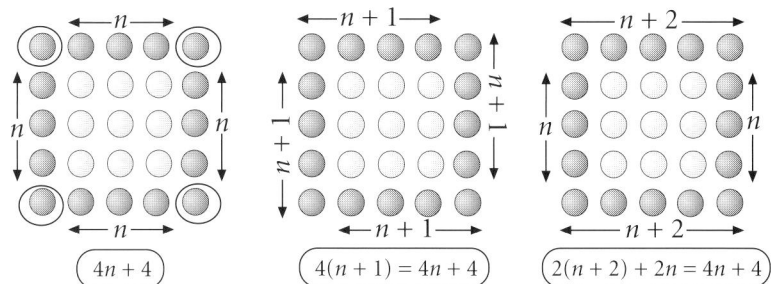

$$4n + 4 \qquad 4(n + 1) = 4n + 4 \qquad 2(n + 2) + 2n = 4n + 4$$

You may need to emphasise the fact that design n has an n by n square of **yellow** Smarties. Hence the number of red Smarties along each edge is not n but $n + 2$.

Ask pupils to use their rules to find the number of red and yellow Smarties needed for design 100, for example.

Ways of seeing further (p 44)

◊ Usually pupils quickly see that to find the number of cans in a stack with, say 100 rows, you need to work out $1 + 2 + 3 + 4 + 5 + \ldots + 99 + 100$ but can't figure out a way to do this quickly or use it to come up with a formula.

Two possible approaches are outlined below.

Approach 1

Imagine the cans ranged to the left so the stacks look like this:

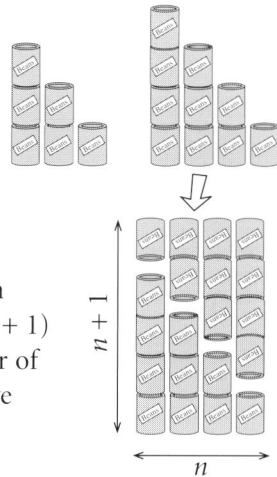

Now take one stack and imagine an identical one rotated 180° and placed on top. You now have a rectangle of tins $(n + 1)$ tins long and n tins wide. So the number of tins is $n(n + 1)/2$ (dividing by 2 as you've got twice as many as you want).

Approach 2

Consider the series and imagine the same series reversed underneath.

$$1 + 2 + 3 + 4 + \ldots + 99 + 100$$
$$100 + 99 + \ldots + 2 + 1.$$

This produces 100 pairs of numbers each adding to 101, so the total is $100 \times \dfrac{101}{2}$ (dividing by 2 as you've got double the total you want). Generalising gives $\dfrac{n(n + 1)}{2}$ as before.

◊ Encourage pupils to look for links between the various rules and contexts in section G.

For example, they may notice that the rule for the stacks of cans at the top of the page and the rule for the mystic roses are very similar: $\dfrac{n(n + 1)}{2}$ and $\dfrac{n(n - 1)}{2}$. They can think about why this is so.

Some pupils may see that the handshakes problem in G3 is essentially the same as the mystic rose problem in G2.

Some may see that earlier work on stacks of cans helps in finding the rule in G4.

G1 Make sure that pupils realise that the n in part (c) refers to the number of rows and not to the number of tins along the bottom row.

Some pupils may notice that the number of tins in a stack with n rows is n^2. Ask them to explain why this will always be the case. The simplest way is probably to imagine each stack divided just to the right or left of the central column of tins. The two 'pieces' can then be rearranged to give an n by n square of tins.

G4 In part (c), pupils are likely to come up with a variety of expressions, for example, $2n(n-1) + n$, $2n^2 - n$ and $\dfrac{4n(n-1)}{2} + n$.

Ask pupils to compare their expressions and try to show that they are equivalent.

Ⓐ **Multiplying** (p 34)

A1 (a) xy (b) $15t$ (c) $6cd$
    (d) $12pq$ (e) m^2 (f) $2a^2$
    (g) $15gh$ (h) $18y^2$ (i) $14n^2$
    (j) $16k^2$

A2 (a) $4gh$ (b) $56m$ (c) $7x^2$
    (d) $8ab$ (e) $5cd$ (f) $3ef$
    (g) $18gh$ (h) $6k^2$ (i) $6a^3$
    (j) $10b^2c$ (k) $6d^3$ (l) $6ef^2$

A3 (a) $2h^3$ (b) $24b^3$ (c) $30s^3t$
    (d) $6g^3$ (e) $15h^2j$ (f) $6m^2n$
    (g) $6a^2b^2$ (h) $8c^3d$ (i) $3f^2g^3$

Cover up

$6a^2$	$4ab$	$8ab^2$	$4a^3$	$6a^2b$
		$8ab$		
$3a^2b$		$8b^3$		$9a^2$
	$6ab$	$10a$		
$4a^2b^3$	$8b^2$	$4ab^2$	$2a^3$	$6a^2b^2$
$12ab$		$5b^2$		

Ⓑ **Dividing** (p 36)

B1 (a) 2 (b) $2q$ (c) n (d) $7g$
    (e) $2b$ (f) $2x$ (g) $4d$ (h) x
    (i) a (j) $5y$ (k) $3k$ (l) $\frac{1}{2}p$

B2 (a) $2y$ (b) x^2 (c) $6ab$

B3 (a) $2q$ (b) $2a$ (c) $4m$

B4 (a) $3n$ (b) 7 (c) $4k$ (d) $3h$
    (e) $2f$ (f) $4e$ (g) 2 (h) $6ab$

B5 (a) $\dfrac{3z}{2}$ (b) $\dfrac{5x}{4}$ (c) $\dfrac{3v}{2w}$
    (d) $\dfrac{2u}{3}$ (e) $\dfrac{4s}{t}$ (f) $\dfrac{3q}{2}$
    (g) $\dfrac{5}{3}$ (h) $\dfrac{4m}{3}$ (i) $\dfrac{7j}{h}$
    (j) $\dfrac{2}{f}$ (k) $\dfrac{10d}{3}$ (l) $\dfrac{1}{3a}$

B6

Ⓒ **Brackets** (p 38)

C1 (a) $x(2x + 9)$, $2x^2 + 9x$
    (b) $3y(2z + 5)$, $6yz + 15y$
    (c) $2m(9m + n)$, $18m^2 + 2mn$
    (d) $3q(p - 5)$, $3pq - 15q$
    (e) $4b(5a - 2)$, $20ab - 8b$
    (f) $4c(10c - 6)$, $40c^2 - 24c$
    (g) $3a(3a + 1)$, $9a^2 + 3a$

C2 A and G   B and D
     C and E   F and H

C3 (a) $2x^2 + 4x$ (b) $y^2 - 5y$

     (c) $15x - 5y$ (d) $6x - x^2$

     (e) $xy + xz$ (f) $6xy + 2x$

     (g) $5x^2 - 6x$ (h) $6x^2 - 10xy$

     (i) $12yz + 8y^2$ (j) $2x^2y + 2xz^2$

     (k) $x^2y - xy^2$ (l) $2x^2y + 6xy^2$

C4 (a) $m + 4$ (b) h (c) $f + g$

     (d) $d - 1$ (e) $2c$ (f) $3a - 2b$

C5 (a) $s(2s - 5) = 2s^2 - 5s$

     (b) $\mathbf{2r}(r + 6) = 2r^2 + \mathbf{12r}$

     (c) $6q(q - \mathbf{2p}) = \mathbf{6q^2} - 12pq$

     (d) $t(2m + n) = \mathbf{2mt} + nt$

     (e) $\mathbf{4k}(2k - 1) = 8k^2 - \mathbf{4k}$

     (f) $hj(2 + \mathbf{j}) = \mathbf{2hj} + hj^2$

     (g) $5f(\mathbf{2g} + \mathbf{1}) = 10fg + 5f$

     (h) $\mathbf{2d}(3e^2 - 1) = \mathbf{6de^2} - 2d$

     (i) $c(\mathbf{c} + \mathbf{5}) = c^2 + 5c$

     (j) $7(\mathbf{a^2} - \mathbf{b}) = 7a^2 - 7b$

D True to form (p 40)

D1 (a) C, D, E, G and H

     (b) E and L   (c) C, I and J

     (d) C, D and E   (e) F, J and K

     (f) B, K and L   (g) A

D2 $6n$ or $6k$ and so on

E Grid totals (p 41)

E1 (a) 105

     (b) The totals for the T-shape are always multiples of 5.

     (c) (i)

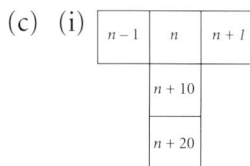

       (ii) $5n + 30$

       (iii) $5n + 30$ can be written $5(n + 6)$ which shows the total is always a multiple of 5.

     (d) Position the 'T' so 34 is the number in the middle of the top row.

E2 (a) The total is always a multiple of 3.

     (b) The total is always a multiple of 2.

     (c) The total is always a multiple of 5 – multiply the centre number by 5 to find the total.

E3 The pupil's shape tha gives a multiple of 6 as its total, for example

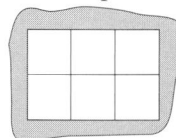

E4 (a) (i)

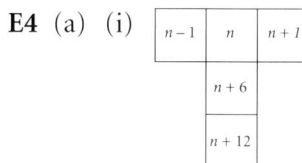

       (ii) $5n + 18$

     (b) The pupil's investigation: the total is always of the form $5n + k$ where k is three times the number of columns.

     (c) $5n + 3m$

F Ways of seeing (p 42)

F1 (a) 24 red Smarties

     (b) Ken $3 \times 9 = 27$

       Gill $10 + 9 + 8 = 27$

       Raj $3 \times 8 + 3 = 27$

     (c) (i) They are all correct.

       (ii) $r = (n + 3) + (n + 2) + (n + 1)$ is Gill's rule.

         $r = 3(n + 1) + 3$ is Raj's rule.

         $r = 3(n + 2)$ is Ken's rule.

         The pupil's explanations

F2 (a) The pupil's drawing of design 4

     (b) (i) 20 yellow Smarties

       (ii) 22 red Smarties

     (c) 110 yellow and 46 red Smarties

     (d) $n(n + 1)$ or $n^2 + n$

     (e) $4n + 6$ or $2(2n + 3)$

     (f) 406 red Smarties

F3 (a) (i) 9 yellow Smarties

(ii) 40 red Smarties

(b) 57 yellow and 904 red Smarties

(c) $4n - 3$

(d) $4n^2 + 4$

(e) $4n^2 + 4$ can be written $4(n^2 + 1)$ so the number of red Smarties is always a multiple of 4.

Ⓖ **Ways of seeing further** (p 44)

G1 (a) 16 tins   (b) 36 tins   (c) n^2

G2 (a) The pupil's mystic roses and numbers of straight lines

(b) 190 lines

(c) $\dfrac{n(n-1)}{2}$ or $\dfrac{n^2 - n}{2}$ or equivalent

G3 (a) 3 handshakes

(b) The pupil's investigation

(c) 4950 handshakes

(d) $\dfrac{n(n-1)}{2}$ or $\dfrac{n^2 - n}{2}$ or equivalent

G4 (a) 45 cubes

(b) (i)   15 cubes

(ii) 435 cubes

(c) The pupil's expression, for example, $2n(n-1) + n$, $2n^2 - n$, $\dfrac{4n(n-1)}{2} + n$.

The pupil's explanation

⋆**G5** (a) There are 5 strands along each edge.

(b) 61 strands

(c) (i)   37 strands

(ii) 127 strands

(d) 1141 strands

(e) The pupil's expression equivalent to $3n^2 - 3n + 1$

One way to find the expression is to split the strands into 3 parallelograms, each one containing n rows of $n - 1$ strands.

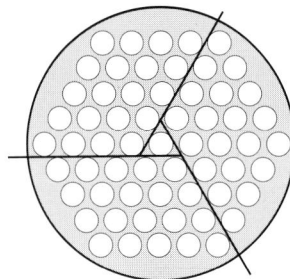

Altogether this gives $3n(n - 1)$ strands for the parallelograms and one extra for the centre strand.

The total is $3n(n - 1) + 1$ which gives $3n^2 - 3n + 1$.

What progress have you made? (p 46)

1 (a) $15ab$ (b) $8k^2$ (c) $12p^2q$

(d) $15r^3$ (e) $9x^2y$

2 (a) $2h$ (b) $6n$

3 (a) $3d$ (b) $\dfrac{h}{3}$ (c) $\dfrac{3}{2x}$

4 (a) $3xy - 3xz$ (b) $6mn + 9m^2$

5 (a) The pupil's totals

(b) If n is in the top left 'cell', the total is $n + (n + 1) + (n + 5) + (n + 6) = 4n + 12 = 4(n + 3)$

(c) 52, 53, 57, 58

Practice booklet

Section A (p 16)

1 (a) ab (b) $3cd$ (c) $4pq$

(d) $12ef$ (e) $20m^2$ (f) $24a^2$

2 (a) $21pq$ (b) $12a^2$ (c) $8e^3$

(d) $30m^2n$ (e) $24b^3$ (f) $10y^3$

(g) $8x^2y$ (h) $6m^2n^2$ (i) $5cd^3$

3 (a) $12abc$ (b) $36a^3$

(c) $3a^2b$ (d) $24x^3$

4 (a) $20a^2b$ (b) $4cd^2$ (c) $60e^2f$

Section B (p 18)

1 (a) $3n$ (b) $4p$ (c) $7x$ (d) $4y$
    (e) $4d$ (f) $4p$ (g) $3b$ (h) $3u$

2 (a) $9s$ (b) $3n$ (c) $4x$ (d) v
    (e) $4a$ (f) $3m$ (g) $\frac{4}{3}v$ (h) $\frac{4}{3}$
    (i) $\frac{2}{3}f$ (j) $2c^2$ (k) $\frac{4}{5}b$ (l) $\frac{v}{3}$

3 (a) $6c$ (b) $4m$ (c) $2v$

Section C (p 19)

1 (a) $2a(3a + 2) = 6a^2 + 4a$
   (b) $m(4m + 3) = 4m^2 + 3m$
   (c) $4x(x - 1) = 4x^2 - 4x$
   (d) $2n(3n - 2) = 6n^2 - 4n$
   (e) $\dfrac{2x(2x + 5)}{2} = 2x^2 + 5x$
   (f) $\dfrac{4y(5y + 6)}{2} = 10y^2 + 12y$

2 (a) $10c + 8c^2$ (b) $6a^2b - 9ab$
   (c) $2y^3 + 3xy^2$ (d) $12d - 6d^2$
   (e) $x^3y - xy^2$ (f) $24z^2 - 6yz$

3 (a) $3 + e$ (b) $2m$
   (c) $3 + 2p$

4 (a) $\mathbf{2x}\,(6x - 2) = 12x^2 - 4x$
   (b) $3m(\mathbf{3} + 2m) = 9m + \mathbf{6m^2}$
   (c) $\mathbf{3}n(5n - 2) = 15n^2 - \mathbf{6n}$
   (d) $\mathbf{7a}\,(\mathbf{3a - 1}) = 21a^2 - 7a$

Sections D and E (p 20)

1 (a) D, G   (b) J
   (c) A, C, E, G, I   (d) A, B, H, I
   (e) C, E

2 (a) 93

   (b) (i)

$n-1$	n	$n+1$
$n+4$		$n+6$
$n+9$		$n+11$

     (ii) $7n + 30$

   (c)

No. of columns	Total
3	$7n + 18$
4	$7n + 24$
5	$7n + 30$
6	$7n + 36$

   (d) $7n + 6m$

Sections F and G (p 21)

1 (a) (i) 20   (ii) 80
   (b) $4n + 4$
   (c) $4n(n + 1) = 4n^2 + 4n$
   (d) Yellow Smarties = 204
     Red Smarties = 10 200

2 (a) 6

   (b)

No. of people	No. of cards
1	0
2	2
3	6
4	12
5	20

   (c) 9900   (d) $n(n - 1)$

*3 (a) 28

   (b) The pupil's three mystic roses and
     number of triangles

No. of points	No. of triangles
4	4
5	10
6	18
7	28
8	40
9	54
10	70

   (c) 340
   (d) $n(n - 2) - n$ or $n^2 - 3n$ or equivalent

6 Area of a circle

Practice booklet pages 22 to 26

A The formula for the area of a circle (p 47)

◊ Diagrams A and B show that the area of a circle is between $2r^2$ and $4r^2$.
The data for the four given circles show that the area is about $3.14r^2$.
Pupils could confirm this for a circle of their own drawn on squared paper.

◊ The diagrams showing a circle cut into sectors and the sectors rearranged should lead to discussion of questions such as

• Roughly what shape is the second diagram?
• Why is it only roughly this shape?
• What is its area, roughly?
• What would happen if the circle were cut into more (thinner) sectors?
• What happens as the number of sectors gets larger and larger?

B Area and circumference (p 49)

The purpose of this section is to help pupils use the right formula – area or circumference.

C Calculating radius given area (p 50)

◊ You may need to remind pupils about the meaning of square root.

D Using exact values (p 51)

◊ Pupils may find it strange at first to be giving results involving the symbol π. Sometimes it is better to do this because π can be 'cancelled out': a good example arises in 'A girdle round about the Earth' (see below).

E Cylinders (p 52)

◊ A reminder about how to find the volume of a prism may be needed.

F Mixed questions (p 53)

A girdle round about the Earth

Ask pupils to have a guess at the answer before they try to work it out.

The answer (about 16 cm) is surprising – even unbelievable!

The calculation shows the advantages of not substituting a numerical value for π at an early stage. It is also helpful to use a symbol, say R, for the radius of the Earth in metres.

The length of the Equator is $2\pi R$ metres.
After the extra 1 metre has been inserted, the length of the band is $(2\pi R + 1)$ metres.
The new radius of the band is therefore $\frac{2\pi R + 1}{2\pi} = R + \frac{1}{2\pi}$ metres.

So the gap between the band and the Earth is $\frac{1}{2\pi}$ metres $= \frac{100}{6.28\ldots}$ cm which is approximately 16 cm.

Notice that R cancels out, so the result is the same whatever the radius of the original circle. The result arises because circumference is proportional to radius, so an additional metre of circumference will always produce the same change in radius.

('I'll put a girdle round about the Earth in forty minutes' – Shakespeare, *A Midsummer Night's Dream*)

A The formula for the area of a circle (p 47)

A1 (a) 113.1 cm²   (b) 176.7 cm²
(c) 18.1 cm²   (d) 98.5 cm²
(e) 2.5 cm²

A2 (a) 1.4 cm   (b) 6.2 cm²

A3 (a) 60.8 cm²   (b) 45.4 cm²
(c) 227.0 cm²   (d) 22.9 cm²
(e) 63.6 cm²   (f) 22.1 cm²

A4 (a) 8.0 cm² (b) 4.5 cm² (c) 3.5 cm²

A5 (a) The pupil's decision
(b) Area of ring (6.4 cm²) is slightly greater than area of green circle (6.2 cm²)

B Area and circumference (p 49)

B1 (a) (i) 13.8 cm (ii) 15.2 cm^2

 (b) (i) 10.1 cm (ii) 8.04 cm^2

 (c) (i) 32.0 cm (ii) 81.7 cm^2

 (d) (i) 27.0 cm (ii) 58.1 cm^2

 (e) (i) 102 cm (ii) 824 cm^2

B2 (a) 141 cm^2 (b) 60.9 cm (c) 464 cm^2

B3 (a) $40\pi = 126$ m

 (b) 6513.3 m^2 = 6510 m^2 to 3 s.f.

B4 264 cm^2

B5 (a) 11.6 cm^2 (b) 21.5%

B6 They are equal. The pupil's explanation

C Calculating radius given area (p 50)

C1 1.95 cm

C2 (a) 2.52 cm (b) 3.99 cm (c) 6.18 cm
 (d) 0.98 cm (e) 0.54 cm (f) 0.69 cm

C3 (a) 3.337790589…

 (b) 21.0 cm

 (c) If $r = 3.3$, circumference = 20.7 cm

C4 23.4 cm

C5

Radius	Diameter	Circumference	Area
–	25.6 cm	80.4 cm	514.7 cm^2
27.9 cm	–	175.3 cm	2445.4 cm^2
11.6 cm	23.1 cm	–	419.4 cm^2
4.5 cm	9.0 cm	28.4 cm	–

C6 17.7 cm

*C7 (a) The pupil's explanation
 (b) 7.25 cm

D Using exact values (p 51)

D1 (a) 10π cm (b) 25π cm^2

D2 (a) (i) 14π cm (ii) 49π cm^2

 (b) (i) 20π cm (ii) 100π cm^2

 (c) (i) 30π cm (ii) 225π cm^2

 (d) (i) 40π cm (ii) 400π cm^2

 (e) (i) 5π cm (ii) 6.25π cm^2

D3 4.5 cm

D4 (a) The pupil's explanation
 (b) $36 - 4.5\pi$ cm^2

D5 (a) $16 - 4\pi$ cm^2 (b) $16 - 4\pi$ cm^2
 (c) $2(\pi - 2)$ cm^2

D6 (a) 2π cm^2 (b) 3π cm^2

D7 $12\pi + 16$ cm^2

D8 (a) $\frac{50}{\pi}$ cm (b) $\frac{10}{\pi}$ cm (c) $\sqrt{\frac{30}{\pi}}$ cm

E Cylinders (p 52)

E1 (a) $2\pi r$ (b) $2\pi rh$
 (c) The pupil's explanation

E2 (a) 185 cm^3 (b) 181 cm^2

E3 (a) 794 cm^3 (b) 514 cm^2

E4

Base radius	Base area	Height	Volume	Surface area
–	50.3 cm^2	–	452 cm^3	327 cm^2
2.76 cm	–	–	240 cm^3	222 cm^2
1.55 cm	7.5 cm^2	–	–	73.2 cm^2
–	28.3 cm^2	2.69 cm	–	107 cm^2

E5 (a) 894 cm^3 (b) 605 cm^2

E6 210π cm^2

*E7 356 cm^3

F Mixed questions (p 53)

F1 7.6 cm

F2 (a) 12π cm (b) 12π cm
 (c) 12π cm^2 (d) 24π cm^2

F3 (a) $80\pi\,\text{cm}^3$ (b) $\dfrac{15}{\pi}\,\text{cm}$

       (c) $\sqrt{\dfrac{20}{\pi}}\,\text{cm}$ (d) $\dfrac{5}{\pi}\,\text{cm}$

*F4 3.7 cm

What progress have you made? (p 54)

1 (a) (i) $78.5\,\text{cm}^2$ (ii) $31.4\,\text{cm}$

    (b) (i) $9.1\,\text{cm}^2$ (ii) $10.7\,\text{cm}$

    (c) (i) $8494.9\,\text{cm}^2$ (ii) $326.7\,\text{cm}$

2 (a) $1.7\,\text{m}$ (b) $12.6\,\text{m}$ (c) $0.2\,\text{m}$

3 (a) $0.2\,\text{m}^2$ (b) $114.9\,\text{m}^2$

    (c) $7162.0\,\text{m}^2$

4 (a) $15\pi\,\text{cm}$ (b) $2.25\pi\,\text{cm}^2$

    (c) $\dfrac{40}{\pi}\,\text{cm}$

5 Volume $= 360\,\text{cm}^3$
    Surface area $= 282\,\text{cm}^2$

Practice booklet

Section A (p 22)

1 (a) $141\,\text{cm}^2$ (b) $60.8\,\text{cm}^2$ (c) $196\,\text{cm}^2$

2 (a) $221.7\ldots\,\text{cm}^2$

    (b) $98.5\ldots\,\text{cm}^2$

    (c) $123.2\ldots\,\text{cm}^2$

3 $198\,\text{m}^2$

4 (a) $121\,\text{cm}^2$ (b) $56.7\,\text{cm}^2$

Section B (p 22)

1 (a) $13.2\,\text{cm}$ (b) $91.6\,\text{cm}^2$

    (c) $55.3\,\text{cm}$ (d) $123\,\text{cm}^2$

2 (a) $20\pi\,\text{m} = 62.8\,\text{m}$

    (b) $800 + 1570.8 = 2370.8\,\text{m}^2$

Section C (p 23)

1 $r = \sqrt{\dfrac{A}{\pi}} = \sqrt{\dfrac{30}{\pi}} = 3.1\,\text{cm}$

    (to the nearest 0.1 cm)

2 (a) $3.8\,\text{cm}$ (b) $5.2\,\text{cm}$ (c) $1.4\,\text{cm}$

    (d) $3.3\,\text{cm}$ (e) $6.4\,\text{cm}$ (f) $2.4\,\text{cm}$

3 (a) $7.42\,\text{cm}$ (b) $23.3\,\text{cm}$

4 (a) $2.54\,\text{cm}$ (b) $2.05\,\text{cm}$

Section D (p 24)

1 (a) $8\pi\,\text{cm}$ (b) $13\pi\,\text{cm}$

    (c) $576\pi\,\text{cm}^2$ (d) $676\pi\,\text{cm}^2$

2 (a) (i) $(5\pi + 10)\,\text{cm}$

       (ii) $12.5\pi\,\text{cm}^2$

    (b) (i) $(6\pi + 26)\,\text{cm}$

       (ii) $(18\pi + 42)\,\text{cm}^2$

    (c) (i) $(16\pi + 32)\,\text{cm}$

       (ii) $(252 - 16\pi)\,\text{cm}^2$

    (d) (i) $18\pi\,\text{cm}$

       (ii) $20\pi\,\text{cm}^2$

Section E (p 24)

1 (a) (i) $1200\,\text{cm}^3$ (ii) $728\,\text{cm}^2$

    (b) (i) $970\,\text{cm}^3$ (ii) $877\,\text{cm}^2$

    (c) (i) $105\,\text{cm}^3$ (ii) $268\,\text{cm}^2$

2 (a) $0.987\,\text{cm}$ (b) $0.156\,\text{cm}$

3 $1.06\,\text{cm}$

Section F (p 25)

1 (a) 5000 square feet

    (b) $5000 - 804.2 = 4195.8$ square feet
       (4200 would be a reasonable answer.)

    (c) 3333.3 square feet (3330 or 3300)

    (c) 2931.2 square feet (2930 or 2900)

2 (a) $196\pi\,\text{cm}^2$ (b) $28\pi\,\text{cm}$

    (c) $192\pi\,\text{cm}^3$ (d) $572\pi\,\text{cm}^2$

3 (a) $903\,\text{cm}^3$ (b) $482\,\text{cm}^2$

4 (a) $25\,\text{cm}$ (b) $20\,\text{cm}$ (c) $3\,\text{cm}$

    (d) $3.5\,\text{cm}$ (e) $49\pi\,\text{cm}^2$ (f) $\frac{1}{16}\pi\,\text{cm}^2$

Review 1 (p 55)

1 (a) 8.2 cm   (b) 55.3 cm
   (c) 4.6 cm   (d) 1.8 cm

2 (a) $x = {}^-6$ (b) $x = 4$ (c) $x = {}^-13$

3 (a) (i) $\frac{1}{2}$ minute   (ii) $1\frac{1}{2}$ minutes
   (b) (i) 1.5 km/min   (ii) 1 km/min
       (iii) 1.25 km/min
   (c) (i) 90 km/h   (ii) 60 km/h
       (iii) 75 km/h

4 (a) $\begin{bmatrix} {}^-3 \\ 6 \end{bmatrix}$ (b) $\begin{bmatrix} {}^-11 \\ 2 \end{bmatrix}$

5 (a) $15x - 6x^2$ (b) $8a^2 - 2ab$
   (c) $p^2q + 3pq^2$ (d) $x^3 - 3x^2y$
   (e) $2ab^2c + 5abc^2$

6 (a) 167 cm^2 (b) 84.9 cm^2 (c) 5.60 cm

7 54 seconds

8 $3n - 20 = \frac{1}{2}(n + 20)$
          $n = 12$

9 (a) 5π cm   (b) 6.25π cm^2
   (c) 25π cm^3 (d) 32.5π cm^2

10 (a) $p = {}^-11$, $q = 2$
    (b) $r = 7$, $s = {}^-4$

11 (a) Line p: gradient 1.5, intercept ${}^-1$
        Line q: gradient ${}^-0.5$, intercept 3
    (b) Line p: $y = 1.5x - 1$
        Line q: $y = {}^-0.5x + 3$ or $y = 3 - \frac{1}{2}x$

12 (a) $7x - 10$ (b) $14a - 3b$
    (c) $p - 6q$ (d) $17x - 6$

13 (a) $x = 4$ (b) $x = 6$ (c) $x = 16\frac{1}{2}$

Mixed questions 1 (Practice booklet p 27)

1 (a) 7600 km to 3 s.f.
   (b) 70 800 km to 3 s.f.

2 (a) $a = 7$ (b) $b = 9.5$
   (c) $c = {}^-\frac{8}{5}$ or ${}^-1.6$ (d) $d = {}^-\frac{3}{2}$ or ${}^-1.5$
   (e) $e = \frac{1}{2}$ (f) $f = 2$

3 9 minutes

4 (a) Inside   (b) $\begin{bmatrix} 2 \\ 4 \end{bmatrix} + \begin{bmatrix} 2 \\ 3 \end{bmatrix}$
   (c) The pupil's journey
   (d) $\begin{bmatrix} {}^-2 \\ 0 \end{bmatrix}$

5 (a) $3a^3b^2$ (b) $\frac{3}{x}$ (c) $\frac{a^2}{4b}$ (d) $\frac{6}{xy}$

6 12π cm

7 The starting number is ${}^-1$ from the
   equation $6 - 4x = 5(x + 3)$.

8 (a) 18°   (b) 20

9 (a) £14 763   (b) 12.125%
   (c) (i) £11 100   (ii) £12 000

10 28 litres

11 (a) $\frac{1}{24}$ (b) $\frac{11}{30}$

⑦ Over to you (p 57)

The intention here is for pupils to try these problems unaided.

Here is one teacher's account of how this material was used:

'I enjoyed doing the problems with the class more than I expected. We did not do all the questions. They appeared to be pitched at the right level; solutions were found but not too quickly.

We discussed each solution as we went along. The pupils did develop a better idea of how to get started, the importance of summarising and organising the information in logical way. We also spent time discussing the presentation of the solutions and I felt this was valuable. It was an opportunity to emphasise the detail needed on strategies used and observations made. Problems were solved in different ways, some equally efficiently.'

A pupil (from a different school) wrote:

'They are good mind teasers, once you start, you tend to want to finish, and it can become quite intense when you can't get the answer – very good and well thought out.'

There are some further comments on individual problems below.

1   Adding three consecutive numbers always gives a multiple of 3.
Adding four consecutive numbers never gives a multiple of 4.

With an odd number of consecutive numbers you get a multiple of the number, but with an even number you don't.

'Question 1 was a good example of how they could use their algebraic skills to predict solutions and prove rules.'

2   Two consecutive numbers will be one even and one odd, so the product will be even and therefore cannot end in 5.

Denoting a number ending in 0 by …0, we have

$$\ldots 0 \times \ldots 1 = \ldots 0 \qquad \ldots 1 \times \ldots 2 = \ldots 2 \qquad \ldots 2 \times \ldots 3 = \ldots 6 \text{ and so on}$$

Going through all possible cases of consecutive numbers, we find that the last digit of the product can be 0, 2 or 6.

3   PQ = 55 miles,
QR = 35 miles,
RP = 25 miles.

This approach by a pupil is very elegant.

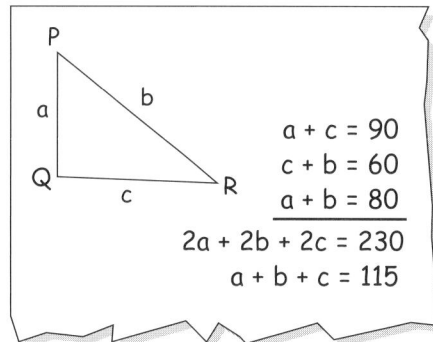

$$a + c = 90$$
$$c + b = 60$$
$$\underline{a + b = 80}$$
$$2a + 2b + 2c = 230$$
$$a + b + c = 115$$

4 A is page 11, B is page 5, C is page 12.

For the sheet from the 24-page newspaper:

p 10	p 15
p 9 (behind)	p 16 (behind)

'This question was set as a homework. Next lesson as a challenge I picked out a double page at random from the local newspaper and told the pupils what the page numbers were. The challenge was to use this information to work out how many pages the paper had that week. This generated a lot of discussion. All pupils got the correct answer.'

5 Square boards:

> Even number of squares on the board: equal numbers of red and white (half the total each).
>
> Odd number on the board: depends on whether top left square is red or white.
> If red, then one more red than white; if white, then other way round.
>
> To find numbers, subtract 1 from total number of squares and divide by 2. This gives smaller number; other is 1 more.

Rectangular boards:

> Again, it depends on the colour of the first square; the rules are the same.

6 (a) 457 + 328 is done as 487 + 325 = 812, which is displayed as 512.

(b) 712 + 453 is displayed as 369.
    In the units place, either 2/6 are interchanged, or 3/7 or 5/9.
    If 2/6 or 5/9, then the tens place doesn't work. So it has to be 3/7.
    This makes the hundreds column work as well.

7 $x = 25°$

'This question proved particularly rich. This was set for a homework and students were invited to write their solutions on the board for class discussion. There were six distinctly different solutions.'

8 130°

9 There are four solutions. A two-way table helps.

	Apple	Banana	Orange
Ann	9	0	8
Bob	2	3	12
Ceri	2	15	0

	Apple	Banana	Orange
Ann	9	0	8
Bob	1	4	12
Ceri	3	14	0

	Apple	Banana	Orange
Ann	10	0	7
Bob	1	3	13
Ceri	2	15	0

	Apple	Banana	Orange
Ann	11	0	6
Bob	1	2	14
Ceri	1	16	0

⑧ Trial and improvement

Essential

Spreadsheet program (section C)

Practice booklet page 29

Ⓐ Introducing the method (p 59)

◊ There is a lot to be said for allowing pupils to record in their own way to start with. Then at a certain point you can ask if their working is such that another person could easily follow it. From this comes the need for some kind of tabulation.

◊ Problem 1 and the first two parts of problem 2 have exact solutions. In the remaining cases an approximate solution can be found more and more accurately. The method of getting a solution correct to a given number of decimal places is dealt with in section B.

◊ The solutions to the problems are as follows.
Problem 1: 17
Problem 2: (a) 5 cm   (b) 6 cm   (c) 5.40312… cm
Problem 3: Each side is 136.568… m.
Problem 4: 9.28… cm by 4.64… cm by 2.32… cm
Contest: 112 cm by 114 cm

Ⓑ Solving equations (p 60)

◊ The number line is a useful image of what is going on and is especially useful when explaining how to get a result correct to, say, two decimal places.

Pupils may not see the need for the last step of working, where the value $x = 1.425$ is tried. They may say it is 'obvious' that x is closer to 1.43 than 1.42 because the corresponding value of $x^3 + 2x^2$ is closer to 7.

However, this reasoning is incorrect, as the following example shows.

Suppose we are trying to solve $x^2 = 2.4$
by trial and improvement.
We know that x lies between 1 and 2.
From the fact that 1^2 is closer to 2.4
than 2^2 is, we might conclude that
the solution is closer to 1 than to 2.
But we would be wrong,
because the solution is $x = 1.549\ldots$,
which is closer to 2.

This happens because the graph of x^2
is curved.

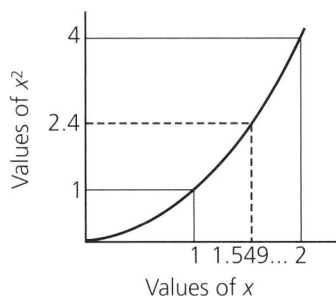

B4 As x increases $8x - x^3$ decreases. This is not usually a problem, so long as pupils keep trying values of x that make the results straddle the target.

$\mathbb{C}$ Using a spreadsheet (p 62)

Spreadsheet

$\Diamond$ After finding that there is a solution between 1.1 and 1.2, it is useful to insert 9 rows between these two values and use them for 1.11, 1.12, … Alternatively you can leave the 1.1 where it is, insert the formula '=B2+0.01' into cell B3 and paste down.

$\mathbb{B}$ Solving equations (p 60)

B1 2.70

B2 1.86

B3 $x^2(x - 1) = 5$ giving $x = 2.12$ to 2 d.p.
Dimensions 2.12 cm by 2.12 cm by 1.12 cm

B4 (a) 8   (b) ¯3   (c) 2.76

B5 (a) 2.37   (b) 0.63

B6 Height = 3.217 cm (base is 6.217 cm)

*__B7__ If the dimensions are x and $(10 - x)$ then $x(10 - x) = 20$, giving $x = 7.24$ to 2 d.p.
So the dimensions are 7.24 cm and 2.76 cm.

$\mathbb{C}$ Using a spreadsheet (p 62)

C1 (a) 1.1642
      (b) ¯3.3914 and ¯1.7729

C2 ¯0.6610, 0.8484 and 1 (exactly)

C3 1.23606797… cm

What progress have you made? (p 62)

1 2.21

Practice booklet

Section B (p 29)

1 (a) $4\,cm^3$ (b) $24\,cm^3$ (c) 1.31 cm
   (1.314 is too small, and 1.315 is too big,
   so x must be less than 1.315)

2 $x = 1.33$

3 2.62 cm

4 (a) 3   (b) 68   (c) $x = 3.21$

5 $x = 2.36$
   so length = 6.36 cm, width = 2.36 cm

⑨ Exploring Pascal's triangle (p 63)

Three different situations are given which lead to Pascal's triangle.
Once the triangle has been generated, it becomes the source of a
variety of other number patterns.

◊ You could introduce each of the three situations and then divide the class
up to work on them.

Buildings

Ask pupils to draw or make the buildings which use 5 square blocks (then
6, 7, …) and to group them by the number on the base. The results can
be recorded in a table.

		Number on base				
		1	2	3	4	…
Number of blocks	1	1				
	2	1	1			
	3	1	2	1		
	4	1	3	3	1	

Two-colour towers

Suppose the two colours are red and yellow. The towers of a given height
have to be grouped by how many, say, red blocks they contain.

		Number of red blocks				
		0	1	2	3	…
Height of tower	1	1	1			
	2	1	2	1		
	3	1	3	3	1	

Routes on a grid

First get agreement on the number of routes in the example given (10).
This may be quite demanding as pupils try to find a way of recording: it
is worth spending some time on this aspect of the task. If, for example, a
route is recorded as UAAUA (A = across, U = up), then the connection
with two-colour towers becomes clear.

Suggest that the grid itself be used to record the number of routes.

◊ Whichever way you get to Pascal's triangle, show it set out like this.

```
        1
       1  1
      1  2  1
     1  3  3  1
    1  4  6  4  1
```

Here are some of the patterns which can be detected.

```
1                          1              Add rows       Triangle numbers
                                                          1
1 +1                       1  1              1            1  1        ↓
1  2  1                    1  2  1           2            1  2  1     ↓
1  3  3 +1                 1  3  3  1         4            1  3  3  1
1  4  6  4  1              1  4  6  4  1      8            1  4  6  4  1
                                            16
```

◊ The first pattern above is the one which defines Pascal's triangle. It can be explained in terms of either the two-colour towers or the routes on a grid.

For example, to get towers of height 5 with 3 reds in them, we can take each tower of height 4 with 2 reds and add an extra red on top, or we can take each tower of height 4 with 3 reds and add an extra yellow on top.

Similarly with the routes:

To get to C …   … either get to A and go across …   … or to B and up.

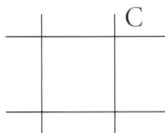

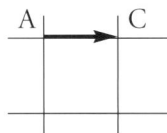

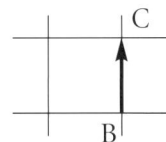

◊ Other patterns which can be explored are given below. Make sure that pupils appreciate that finding a pattern is not the same as proving that it will always be true. Explanations are given below. Pupils would not be expected to discover these for themselves, but some may be able to follow them and thus begin to appreciate the difference between confirming individual cases and proving a general rule.

```
1
1  1              1 + 3 + 6 + 10 = 20
1  2 |1|
1  3 |3|  1
1  4 |6|  4  1
1  5 |10| 10  5  1
1  6  15 |20| 15  6  1
```

Explanation:

```
1
1  1
1  2  1
1  3  3  1
1  4  6  4  1
1  5  10  10  5  1
1  6  15  (20)  15  6  1
```

20 (in circle)   = 10 + 10
= 10 + 6 + 4
= 10 + 6 + 3 + 1
= 10 + 6 + 3 + 1

Fibonacci

1 1 2 3 5 8

1
1 1
1 2 1
1 3 3 1
1 4 6 4 1
1 5 10 10 5 1

Explanation:

0 1
0 1 1
0 1 △2 ①
0 △1 ③ 3 1
0 △1 4 6 4 1
0 ① 4 6 4 1
0 1 5 10 10 5 1

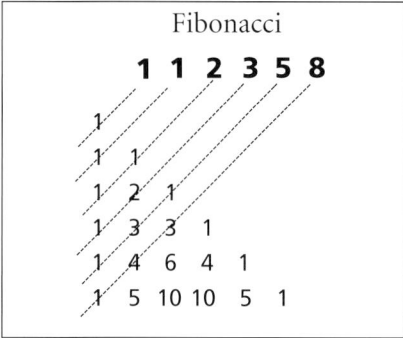

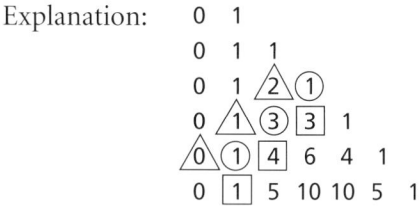

Each ☐ number is the sum of a ◯ number and a △ number.

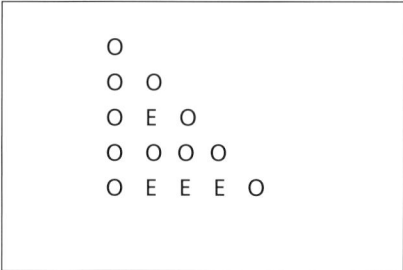

O
O O
O E O
O O O O
O E E E O

This triangle can be generated by using O + O = E, O + E = O and E + E = E (O = odd, E = even). Higher attainers could look at the proportions of odd and even numbers after n rows.

They could also investigate the diagram obtained on squared paper by shading 'O' squares and leaving 'E' squares blank:

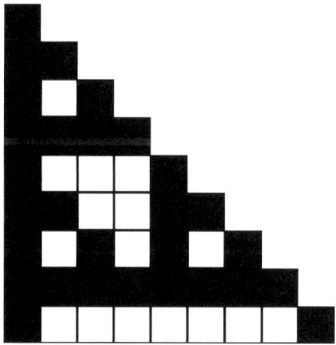

⑩ Calculating probabilities

This unit starts with multiplication of fractions and then applies this to calculating probabilities where independent outcomes are involved. It also introduces tree diagrams.

Practice booklet pages 30 to 34

Ⓐ Fractions of fractions (p 64)

Fractions of fractions are approached pictorially. The link with multiplication is made in the next section.

◊ After the introduction, questions A1–A3 could be done orally. A comparison of pupils' answers to question A4 would be valuable.

Ⓑ Multiplying fractions (p 66)

In older books on arithmetic you will often read that 'of means multiply', but it should be the other way round. We first make sense of fractions of fractions and then define multiplication as this operation.

Ⓒ Traffic flows (p 67)

This is an application of multiplying fractions which has obvious connections with tree diagrams (see section E).

Ⓓ Independent events (p 68)

The shaded square diagrams used in section A are used to explain the multiplication rule.

Ⓔ Tree diagrams (p 70)

◊ It helps to start by using a tree diagram to show possible outcomes, without probabilities. Each branching ⟨ represents a choice.

You could draw tree diagrams for, say, choices of first and second course in a meal.

◊ The next step is to introduce probabilities, as in the example in the pupil's book. The 'traffic flow' analogy may help pupils to get a feel for the way in which the probability reduces.

◊ Addition of fractions is needed in questions E5 to E10. Section F is for revision of this, if necessary.

𝔽 Adding fractions: a reminder (p 72)

𝔸 Fractions of fractions (p 64)

A1 (a) $\frac{1}{4}$ of $\frac{1}{6} = \frac{1}{24}$ (b) $\frac{1}{3}$ of $\frac{1}{6} = \frac{1}{18}$

(c) $\frac{1}{5}$ of $\frac{1}{6} = \frac{1}{30}$

A2 $\frac{6}{12}$ or $\frac{1}{2}$

A3 $\frac{9}{20}$

A4 The pupil's diagrams and fraction statements

A5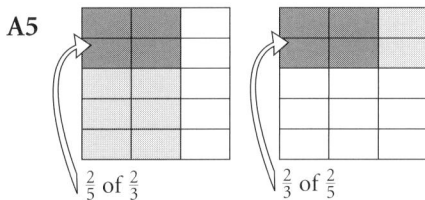

$\frac{2}{5}$ of $\frac{2}{3}$ $\frac{2}{3}$ of $\frac{2}{5}$

𝔹 Multiplying fractions (p 66)

B1 (a) $\frac{1}{6}$ (b) $\frac{1}{12}$ (c) $\frac{1}{18}$

(d) $\frac{1}{12}$ (e) $\frac{1}{10}$

B2 (a) $\frac{3}{10}$ (b) $\frac{9}{20}$ (c) $\frac{3}{20}$ (d) $\frac{8}{15}$

(e) $\frac{9}{40}$ (f) $\frac{4}{25}$ (g) $\frac{4}{15}$

B3 (a) $\frac{6}{12}$ or $\frac{1}{2}$ (b) $\frac{6}{40}$ or $\frac{3}{20}$ (c) $\frac{6}{20}$ or $\frac{3}{10}$

(d) $\frac{10}{18}$ or $\frac{5}{9}$ (e) $\frac{15}{24}$ or $\frac{5}{8}$

B4 $\frac{2}{3} \times \frac{4}{5} = \frac{2 \times 4}{3 \times 5} = \frac{8}{15}$

B5 (a) $\frac{1}{4}$ (b) $\frac{9}{32}$ (c) $\frac{4}{15}$

(d) $\frac{15}{32}$ (e) $\frac{1}{2}$

ℂ Traffic flows (p 67)

C1 (a) $\frac{1}{6}$ (b) $\frac{1}{6}$ (c) $\frac{1}{2}$

C2 (a) $\frac{1}{12}$ (b) $\frac{1}{6}$ (c) $\frac{3}{8}$ (d) $\frac{3}{8}$

C3 (a) (i) $\frac{1}{20}$ (ii) $\frac{3}{20}$ (iii) $\frac{8}{15}$ (iv) $\frac{4}{15}$

(b) D 3, E 9, F 32, G 16

𝔻 Independent events (p 68)

D1 (a) $\frac{1}{6}$ (b) $\frac{1}{2}$ (c) $\frac{1}{4}$

D2 $\frac{1}{30}$

D3 (a) $\frac{4}{5}$ (b) $\frac{1}{3}$ (c) $\frac{4}{15}$ (d) Multiply

D4 $\frac{1}{5}$

D5 $\frac{5}{48}$

𝔼 Tree diagrams (p 70)

E1 (a) $\frac{3}{20}$ (b) $\frac{4}{20} = \frac{1}{5}$ (c) $\frac{12}{20} = \frac{3}{5}$

E2

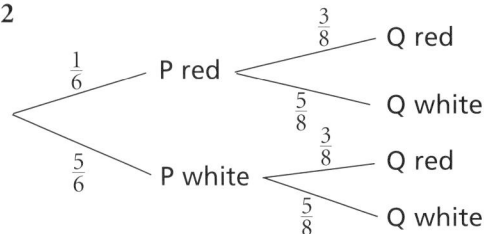

(a) $\frac{3}{48} = \frac{1}{16}$ (b) $\frac{25}{48}$ (c) $\frac{5}{48}$ (d) $\frac{15}{48} = \frac{5}{16}$

E3 (a)

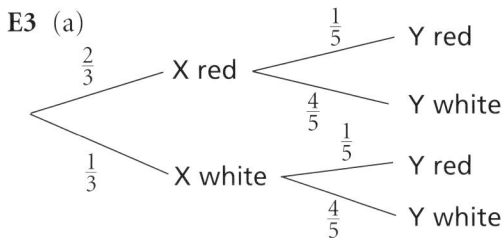

(b) X red Y red: probability $\frac{2}{15}$

X red Y white: probability $\frac{8}{15}$

X white Y red: probability $\frac{1}{15}$

X white Y white: probability $\frac{4}{15}$

E4 (a)

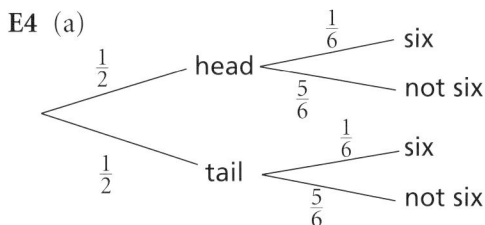

(b) (i) $\frac{1}{12}$ (ii) $\frac{5}{12}$ (iii) $\frac{1}{12}$ (iv) $\frac{5}{12}$

E5 $\frac{28}{48} = \frac{7}{12}$

E6 $\frac{6}{15} = \frac{2}{5}$

E7 $\frac{20}{48} = \frac{5}{12}$

E8 $\frac{9}{15} = \frac{3}{5}$

E9 (a) $\frac{7}{15}$ (b) $\frac{8}{15}$

E10 (a) $\frac{19}{35}$ (b) $\frac{16}{35}$

𝔽 **Adding fractions: a reminder** (p 72)

F1 (a) $\frac{8}{15}$ (b) $\frac{7}{12}$ (c) $\frac{11}{15}$

(d) $\frac{31}{40}$ (e) $\frac{7}{15}$

F2 (a) $\frac{9}{20}$ (b) $\frac{11}{20}$

F3 (a) $\frac{19}{24}$ (b) $\frac{5}{24}$

What progress have you made? (p 72)

1 (a) $\frac{3}{20}$ (b) $\frac{3}{10}$

2 (a) $\frac{3}{15} = \frac{1}{5}$ (b) $\frac{4}{15}$

3 $\frac{8}{15}$

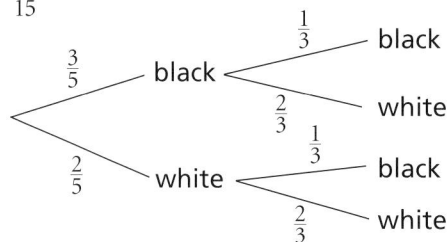

Practice booklet

Section A (p 30)

1 (a) $\frac{1}{6}$ (b) $\frac{1}{10}$ (c) $\frac{1}{20}$ (d) $\frac{1}{32}$

2

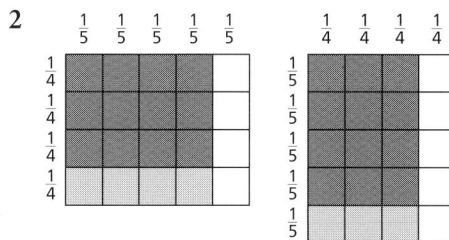

Section B (p 30)

1 (a) $\frac{1}{24}$ (b) $\frac{1}{30}$ (c) $\frac{5}{24}$

2 (a) $\frac{10}{24} = \frac{5}{12}$ (b) $\frac{3}{24} = \frac{1}{8}$

(c) $\frac{12}{20} = \frac{3}{5}$ (d) $\frac{14}{24} = \frac{7}{12}$

3 (a) $\frac{12}{36} = \frac{1}{3}$ (b) $\frac{45}{60} = \frac{3}{4}$

(c) $\frac{24}{36} = \frac{2}{3}$ (d) $\frac{20}{30} = \frac{2}{3}$

4 (a) $\frac{10}{45} = \frac{2}{9}$ (b) $\frac{6}{15} = \frac{2}{5}$

(c) $\frac{35}{60} = \frac{7}{12}$ (d) $\frac{21}{80}$

5 (a) $\frac{12}{60} = \frac{1}{5}$ (b) $\frac{6}{36} = \frac{1}{6}$

(c) $\frac{24}{72} = \frac{1}{3}$ (d) $\frac{15}{30} = \frac{1}{2}$

6 (a) $\frac{9}{30} = \frac{3}{10}$ (b) $\frac{15}{48} = \frac{5}{16}$

(c) $\frac{7}{18}$ (d) $\frac{3}{21} = \frac{1}{7}$

Section C (p 31)

1 (a)(i) $\frac{1}{8}$ (ii) $\frac{1}{8}$ (iii) $\frac{1}{4}$ (iv) $\frac{1}{2}$

(b) The pupil's check

2 (a) $\frac{1}{24}$ (b) $\frac{5}{24}$ (c) $\frac{1}{2}$ (d) $\frac{1}{4}$

3 (a)(i) $\frac{2}{15}$ (ii) $\frac{1}{5}$ (iii) $\frac{1}{6}$ (iv) $\frac{1}{2}$

(b) R 8,   S 12,   T 10,   U 30

Section D (p 32)

1 (a) $\frac{1}{2}$ (b) $\frac{2}{3}$ (c) $\frac{1}{2} \times \frac{2}{3} = \frac{1}{3}$

2 (a) $\frac{2}{5} \times \frac{1}{3} = \frac{2}{15}$ (b) $\frac{2}{5} \times \frac{2}{3} = \frac{4}{15}$
   (c) $\frac{3}{5} \times \frac{1}{3} = \frac{1}{5}$

3 $\frac{1}{3} \times \frac{1}{4} = \frac{1}{12}$

4 $\frac{4}{5} \times \frac{1}{3} = \frac{4}{15}$

Section E (p 33)

1 (a) $\frac{1}{24}$ (b) $\frac{5}{24}$ (c) $\frac{15}{24} = \frac{5}{8}$

2 (a)

(b) red red $\frac{1}{6}$ red white $\frac{1}{6}$
    white red $\frac{1}{3}$ white white $\frac{1}{3}$

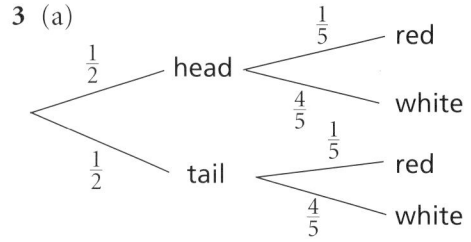

3 (a)

(b) (i) $\frac{1}{10}$ (ii) $\frac{2}{5}$ (iii) $\frac{1}{10}$ (iv) $\frac{2}{5}$

4 (a) $\frac{1}{2} + \frac{1}{12} = \frac{7}{12}$ (b) $\frac{1}{6} + \frac{1}{4} = \frac{5}{12}$

Section F (p 34)

1 (a) $\frac{1}{4} = \frac{5}{20}$ $\frac{1}{5} = \frac{4}{20}$ $\frac{1}{4} + \frac{1}{5} = \frac{5}{20} + \frac{4}{20} = \frac{9}{20}$

   (b) $\frac{2}{3} = \frac{8}{12}$ $\frac{3}{4} = \frac{9}{12}$
       $\frac{2}{3} + \frac{3}{4} = \frac{8}{12} + \frac{9}{12} = \frac{17}{12} = 1\frac{5}{12}$

2 (a) $\frac{13}{40}$ (b) $\frac{5}{12}$ (c) $\frac{13}{15}$ (d) $\frac{33}{40}$

3 (a) $\frac{7}{12}$ (b) $\frac{5}{12}$

4 (a) $\frac{11}{15}$ (b) $\frac{4}{15}$

5 (a) $\frac{13}{40}$ (b) $\frac{17}{40}$

⑪ Linear equations and graphs

Optional

Graph-plotting computer program
or graphical calculator

Practice booklet pages 35 to 37

Ⓐ Equations, tables, graphs (p 73)

◊ Use your introduction to ensure that pupils realise that we can move between tables, coordinates, equations and graphs.

Emphasise that there is more than one way to write the equation that connects the values in a table.

Ⓑ Different forms for the equation of a line (p 75)

Ⓒ Parallel lines (p 76)

Ⓓ Simultaneous equations (p 78)

◊ The most important idea to get over is that the point of intersection of the graphs of two equations has coordinates which fit both equations.

A graph-plotting program that can be used in slow motion, or a graphical calculator, can help to show this. With the graph of the first equation plotted, the graph of the second can be plotted slowly across the screen, with the coordinates of the point being plotted shown. Pupils can see that each value of x leads to different y values for the two equations, except at the intersection, where the y values are identical.

Ⓐ Equations, tables, graphs (p 73)

A1 B, C, E, F

A2 (a) $y = x + 2$ (b) $x = y - 2$

(c) $y - x = 2$ or $x - y = {}^-2$

A3 (a) $y = 6 - x$, $x = 6 - y$, $x + y = 6$

(b) $y = 3x + 3$, $y - 3x = 3$, $x = \dfrac{y-3}{3}$,
$3x - y = {}^-3$

(c) $y = x - 3$, $x = y + 3$, $x - y = 3$,
$y - x = {}^-3$

(d) $y = {}^-x$, $x = {}^-y$, $x + y = 0$

A4 Two out of each of these lists:

(a) $b = 2a$, $a = \frac{1}{2}b$, $b - 2a = 0$,
$2a - b = 0$

(b) $b = {}^-3a$, $a = \frac{-1}{3}b$, $3a + b = 0$

(c) $b = \frac{1}{2}a$, $a = 2b$, $a - 2b = 0$,
$2b - a = 0$

(d) $b = \frac{1}{2}(a - 1)$, $b = \frac{1}{2}a - \frac{1}{2}$, $a = 2b + 1$,
$a - 2b = 1$, $2b - a = {}^-1$

A5

x	y
0	$^-1$
1	2
3	**8**
6	17

A6 (a)

x	y
$^-4$	0
$^-3$	$\frac{1}{2}$
$^-2$	1
0	2

(b)

g	h
0	**10**
3	**7**
9	1
10	0

(c)

s	t
0	**3**
$^-4$	**6**
4	0
8	$^-3$

(d)

x	y
4	**2**
5	1
8	$^-2$
12	$^-6$

A7 (a)

x	y
1	0
1	3
1	4
1	$^-3$

(b)

x	y
0	**2**
$^-3$	**2**
1	**2**
4	**2**

(c)

x	y
$^-4$	4
2	$^-2$
$^-1$	1
0	**0**

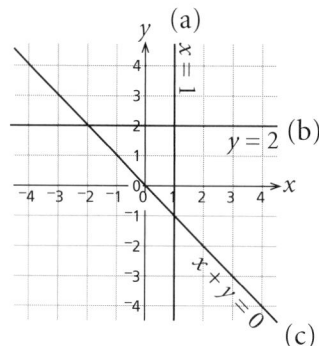

Ⓑ Different forms for the equation of a line (p 75)

B1

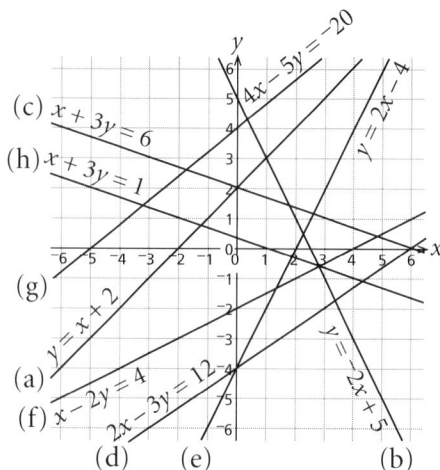

*B2 (a) $a = 4$, $b = 5$

(b) (i) $3x + 7y = 21$

(ii) $2x - 4y = 8$ or $^-2x + 4y = {}^-8$
or $x - 2y = 4$

Ⓒ Parallel lines (p 76)

C1 (a) $(0, 4)$ (b) $y = 2x + 1$

C2 (a) $(0, 3)$ (b) $y = \frac{-1}{2}x - 1$

C3 (a) $(0, 4)$ (b) $y = 3x - 3$
(c) $y = \frac{1}{2}x + 4$

C4 (a) 2 (b) $y = 2x + 1$
(c) $y = 2x - 3$ (d) $(5, 11)$ and $(5, 7)$

C5 (a) $\frac{-1}{3}$ (b) $y = \frac{-1}{3}x + 2$
(c) $y = \frac{-1}{3}x - 3$ (d) $(^-9, 0)$

*C6 (a) $y = 1\frac{1}{2}x + 6$ (b) $y = {}^-1\frac{1}{2}x + 6$
(c) $y = {}^-1\frac{1}{2}x - 6$

$\star$**C7** (a) $y = \frac{-2}{3}x + 6$ (b) $y = \frac{1}{3}x - 3$

(c) $y = \frac{-2}{3}x + 10$ (d) $y = \frac{1}{3}x + 6$

D **Simultaneous equations** (p 78)

D1 (a), (b)

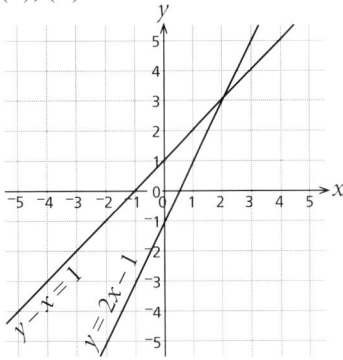

(c) $(2, 3)$ and check

D2

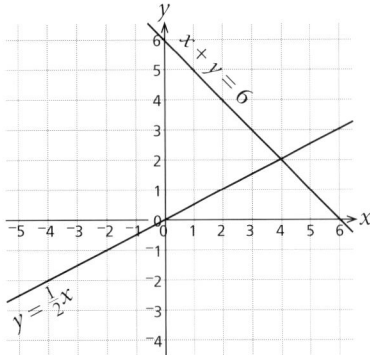

The point $(4, 2)$ fits both equations.

D3 (a)

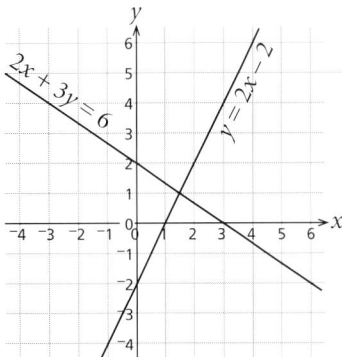

(b) $(1\frac{1}{2}, 1)$ and check

D4 (a)

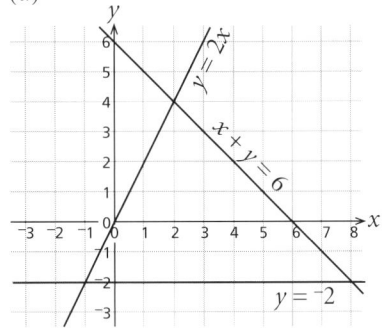

(b) $(^-1, ^-2)$, $(2, 4)$, $(8, ^-2)$

(c) The area is 27 (square units).

D5 (a) $x = 0.8$, $y = 1.8$

(b) $x = 0$, $y = 1$

(c) $x = 2.5$, $y = ^-0.75$

(d) $x = 1.2$, $y = ^-1.4$

D6 (a)

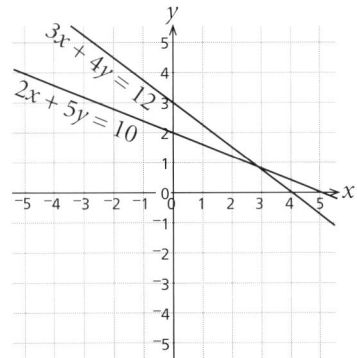

(b) $x = 2.9$, $y = 0.9$ (approximately)

D7 (a)

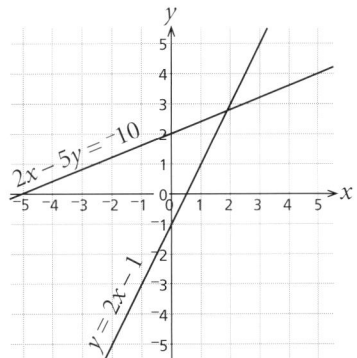

(b) $x = 1.9$, $y = 2.8$ (approximately)

D8 (a) The lines are parallel;
the equations have no solution.

   (b) The lines are identical;
the equations have infinitely many
solutions.

What progress have you made? (p 80)

1

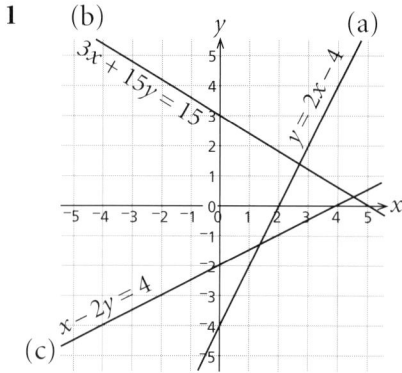

2 (a) $y = \frac{-1}{3}x - 2$

   (b) $(^-6, 0)$

3

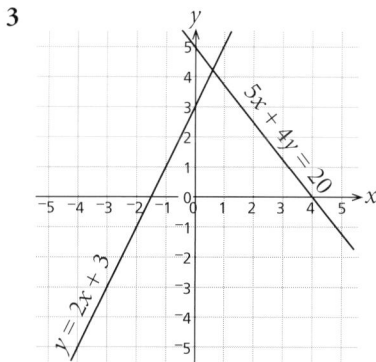

$x = 0.6, y = 4.2$ (approximately)

Practice booklet

Sections A and B (p 35)

1 (a) $x + y = 4$ and equivalent

   (b) $y = 2x - 3$ and equivalent

   (c) $y = \frac{1}{2}x + 1$ and equivalent

2 (a)

x	y
1	0
7	3
5	**2**
0	$-\frac{1}{2}$

   (b)

x	y
0	**5**
4	1
3	**2**
6	$^-1$

   (c)

x	y
0	**4**
10	0
5	2
$2\frac{1}{2}$	3

3

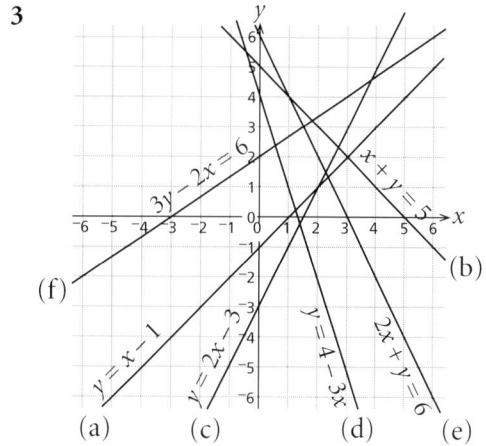

Section C (p 35)

1 (a) $(0, {}^-2)$ (b) $y = 3x + 3$

2 (a) $(0, 3)$ (b) $y = 2x + 15$

   (c) $y = 3 - \frac{1}{2}x$

3 (a) 3   (b) $y = 3x - 2$

   (c) $y = 3x + 3$

4 (a) $(0, 4)$ (b) $^-2$

   (c) $y = 4 - 2x$ (d) $y = \frac{1}{2}x$

   (e) R $(6, 3)$, S $(6, 7)$

Section D (p 37)

1 (a)

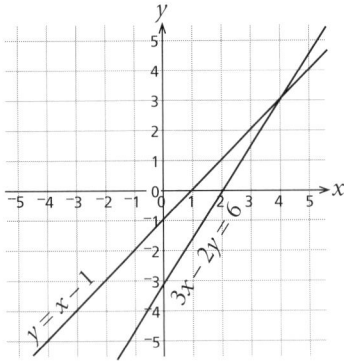

(b) (4, 3)

2 ($^-$2, 1)

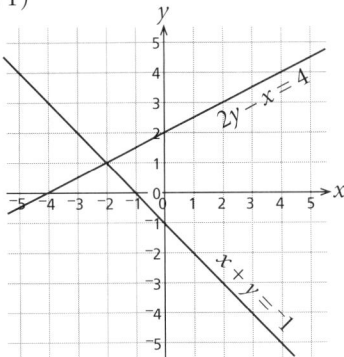

3 (a), (c)

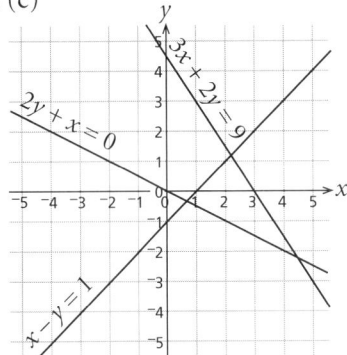

(b) (2.2, 1.2)

(d), (e) (4.5, $^-$2.25) for $3x + 2y = 9$
and $2y + x = 0$ and (0.6, $^-$0.3) for
$x - y = 1$ and $2y + x = 0$

4 (a) A and D   (b) A and C

    (c) B and D

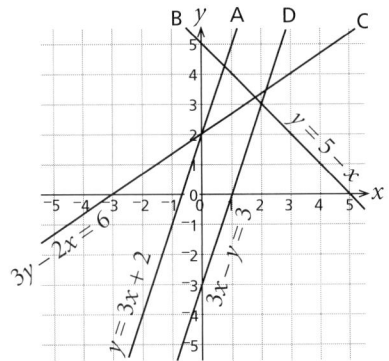

(d) A and B (0.75, 4.25)
$y = 3x + 2$ and $y = 5 - x$

B and C (1.8, 3.2)
$y = 5 - x$ and $3y - 2x = 6$

C and D (2.1, 3.4)
$3y - 2x = 6$ and $3x - y = 3$

⑫ Using and misusing statistics

Practice booklet page 38

A Misleading charts and pictures (p 81)

◊ Whether a chart 'misleads' or not is to some extent subjective. A chart which distorts the relative values of quantities is misleading. One which uses a false origin correctly but tries to draw the reader's attention away from it by emphasising other features may mislead only the unwary (like a tempting offer with almost illegible 'small print').

◊ The first chart ('YUK Computers') is only misleading to someone who does not notice that the vertical scale starts at 2000. It would be better if this were shown by a jagged axis and broken bars.

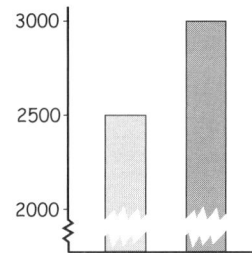

◊ The second chart distorts by making the second bar wider as well as higher than the first.

◊ Showing the pie chart at an angle makes it harder to tell the relative sizes of the slices, taking away the main reason for using such a chart. (Teachers shown this chart have come up with a surprising range of answers. The actual proportions shown are meat 30%, gravy 30% and crust 40%.)

◊ In the lottery chart the heights of the flags are in proportion to the amounts, but the different widths mislead. (A more subtle point is the lack of any reference to the different sizes of the countries shown and the size of the lottery contributions in relation to the national incomes.)

B Problems arising in collecting data (p 84)

Defining categories

The person presenting data and the person reading it may have different understandings of the categories involved. Unless we know what counts as a 'science subject', the statement about students is not clear.

Avoiding bias

Bias may be difficult to eliminate. For example, the people who are willing to complete written questionnaires may not be representative of people as a whole. Also, customers who have a complaint may be more likely to give their views than those who are satisfied.

A Misleading charts and pictures (p 81)

A1 (a) The scale does not start at zero –
looks like Japan has twice the
percentage of the UK, for example.

(b) The labels here are part of the bars.
This makes them look
disproportionate. The bars actually
start some 27 mm from the left.

A2 (a) The vertical scale goes up in very
uneven steps. The pictures also have
disproportionate area.

(b) A classic example where the height
this year may be 1.5 times last year's
but the area is $1.5^2 = 2.25$ times
bigger.

A3 Putting a slice at the front exaggerates its
significance. 'Leisure' shows the effect of
different positions well.

A4 Bembo appears to have had a more rapid
rise but in fact both companies' sales have
increased by a factor of about 4.

A5 The first graph does not start at zero.
The second does not go up to 100%.

B Problems arising in collecting data (p 84)

B1 It is not clear where one age group ends
and another begins, e.g. under 10, 11–20,
21–30, 31–40, over 40

B2 Some programmes could arguably be put
into more than one category, e.g. a
documentary on child abuse – factual or
current affairs?
Some types of programme have no
obvious category, e.g. quiz shows.

B3 The definitions of 'young' and of 'pop'

B4 (a) Many people do not organise
holidays through travel agents.

(b) This is biased to the type of people
who listen to that radio station and
have the time and interest to phone
in.

(c) This is an unrepresentative sample.
Many people rarely, if ever, visit a
bank.

(d) Some people avoid shopping on a
rainy day. Some people rarely shop
in town centres.

What progress have you made? (p 85)

1 The vertical scale does not start at zero.
Years are not evenly spaced.

2 Many people do not visit libraries.
Some visit libraries in the evening.

Practice booklet

Sections A and B (p 38)

1 (a) C

(b) In A, the bar for 2002 is wider than
that for 2001. In B, the vertical scale
starts at 8000; if this is not spotted,
then the 2002 looks four times as
high as the 2001 bar.

2 It is not clear what 'fan' means.
Does 'people of Holby' include
all ages of resident?

3 It would be unlikely that many
questionnaires would be filled in. They
could be filled in by the supporters of the
opposing team. (And other reasons)

⓭ Pythagoras's theorem

Essential	Optional
Square dotty paper	Squared paper
Set square	OHP
Practice booklet pages 39 to 41	

A Tilted squares (p 86)

As well as providing revision of area work, this section prepares for the investigation in section B.

Square dotty paper

B Squares on right-angled triangles (p 87)

Square dotty paper
Set square

◊ Each pupil can be allocated triangles of different dimensions to draw and find the areas of the squares. Then groups of pupils or the whole class can pool their results into a table, from which the general result should emerge.

B6 This demonstrates that Pythagoras actually works!

B7 In some parts of this question the hypotenuse has to be found and in others a side next to the right angle has to be found. The reason for mixing is to require pupils to think about what they are doing, not just plug values into a formula. You may find that you need to spend some time working with additional examples of these types of questions.

C **Using square roots** (p 90)

Set square

D **Using Pythagoras** (p 91)

Optional: Squared paper

E **Proving Pythagoras** (p 93)

Although placed after section D, you could do this demonstration as a break at some point while pupils are working on section D.

Scissors Optional: OHP

◊ This is best demonstrated on an OHP.

The triangles can be arranged on the square
like this ...   ... or like this.

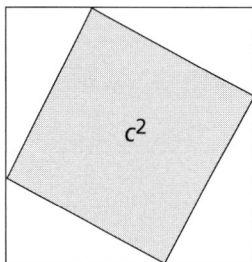

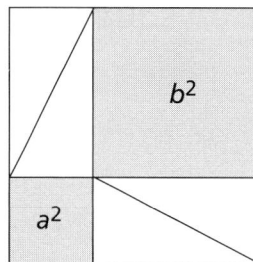

The key points to come out of discussion of the demonstration are

• Since the triangles cover the same area on the pink square in both arrangements, the same amount of pink must be showing in both cases.

• This shows that $c^2 = a^2 + b^2$.

• This result doesn't depend on a and b having particular values. You can change the dimensions of the triangle so long as as you also change the large square so that its sides are still $a + b$.

Since this works for any right-angled triangle it is an acceptable general proof (even though it is visual and informal). There are many valid proofs of Pythagoras's theorem. This one is perhaps the simplest.

A Tilted squares (p 86)

A1 (a) 20 (b) 26 (c) 18 (d) 34

A2 The areas are 2, 5, 10, 17, 26, …
The area of the nth square is $n^2 + 1$.

B Squares on right-angled triangles (p 87)

B1 (a) $12\,\text{cm}^2$ (b) $12\,\text{cm}^2$ (c) $8\,\text{cm}^2$
(d) $13\,\text{cm}^2$ (e) $33\,\text{cm}^2$ (f) $26\,\text{cm}^2$

B2 (a) $64\,\text{cm}^2$ (b) $8\,\text{cm}$

B3 $32\,\text{cm}^2$

B4 (a) $6\,\text{cm}$ (b) $33\,\text{cm}^2$ (c) $185\,\text{cm}^2$
(d) $7\,\text{cm}$ (e) $4\,\text{cm}$ (f) $5\,\text{cm}$

B5 (a) $52\,\text{m}^2$ (b) $7\,\text{m}$ (c) $5\,\text{m}$
(d) $77\,\text{m}^2$

B6 (a) $17\,\text{cm}$ (b) The pupil's check

B7 (a) $9\,\text{cm}$ (b) $12\,\text{cm}$ (c) $10\,\text{cm}$
(d) $5\,\text{cm}$ (e) $13\,\text{cm}$ (f) $26\,\text{cm}$

C Using square roots (p 90)

C1 (a) $23.0\,\text{cm}$ (b) $17.3\,\text{cm}$ (c) $20.8\,\text{cm}$
(d) $22.8\,\text{cm}$ (e) $30.5\,\text{cm}$ (f) $15.8\,\text{cm}$

C2 (a) $14.9\,\text{cm}$ (b) The pupil's check

D Using Pythagoras (p 91)

D1 (a) $24.4\,\text{cm}$ (to nearest 0.1 cm)
(b) $34.4\,\text{cm}$ (to nearest 0.1 cm)

D2 Results for the pupil's book

D3 About 26 metres shorter

D4 (a) $5.83\ (\sqrt{34})\ \text{cm}$
(b) The pupil's check
(c) $13\,\text{cm}$

D5 (a) All sides 7.07… units ($\sqrt{50}$ units)
(b) It is a rhombus.

D6 The shorter sides are each 3.16… ($\sqrt{10}$).
The longer sides are each 7.07… ($\sqrt{50}$).
It is a kite.

D7 $AC = \sqrt{754}$
$AB = \sqrt{821} = 28.7$ (to 1 d.p.)
$BC = \sqrt{845} = 29.1$ (to 1 d.p.)
$CA = \sqrt{754} = 27.5$ (to 1 d.p.)
A and C are closest together.
B and C are furthest apart.

D8 (a) $5.66\,\text{cm}$ (b) $4.47\,\text{cm}$ (c) $9.49\,\text{cm}$
(d) $8.06\,\text{cm}$ (e) $5.00\,\text{cm}$ (f) $10.44\,\text{cm}$
(g) $7.28\,\text{cm}$ (h) $6.40\,\text{cm}$ (i) $10.30\,\text{cm}$

D9 About $19\,\text{m}$

D10 $32.2\,\text{km}$

D11 (a) $0.71\,\text{m}$ (to 2 d.p.) (b) $0.5\,\text{m}^2$

D12 5 units

D13 $10.95\,\text{cm}$ (to 2 d.p.)

*D14 The longest side is 24 cm, so this is the
only side to consider as a possible
hypotenuse. If the largest angle were 90°,
the longest side would be $\sqrt{19^2 + 17^2} = \sqrt{650} = 25.5\,\text{cm}$.
But the longest side is less than this, so
the largest angle is less than 90°.
So all three angles are acute.

What progress have you made? (p 94)

1 (a) $10.4\,\text{m}$ (b) $5.9\,\text{m}$ (c) $12.7\,\text{m}$

2 $2.56\,\text{cm}$, $1.84\,\text{cm}$, $2.56\,\text{cm}$

3 (a) $5.0\,\text{cm}$ (b) $7.2\,\text{cm}$ and $9.8\,\text{cm}$

Practice booklet

Section B (p 39)

1 (a) $11\,\text{cm}^2$ (b) $21\,\text{cm}^2$ (c) $2\,\text{cm}^2$

2 (a) $59\,\text{cm}^2$ (b) $5\,\text{cm}$ (c) $19\,\text{cm}^2$

3 (a) $5\,\text{cm}$ (b) $60\,\text{cm}$ (c) $3\,\text{cm}$
(d) $100\,\text{cm}$ (e) $2\,\text{cm}$ (f) $6\,\text{cm}$

Section C (p 40)

1 (a) 3.1 cm   (b) 18.2 cm   (c) 40.0 m

   (d) 88.7 cm   (e) 7.9 m   (f) 9.6 mm

2 (a) The pupil's triangle XYZ drawn
accurately

   (b) YZ measured, approximately 4.3 cm

   (c) 4.3 cm

Section D (p 40)

1 A 12.6 cm,   B 12.5 cm,   C 12.7 cm
C has the longest diagonal.

2 1.3 m

3 $AB = \sqrt{605} = 24.6$
$BC = \sqrt{637} = 25.2$
$AC = \sqrt{626} = 25.0$

   (a) BC is longest.   (b) AB is shortest.

4 Both the longer sides $= \sqrt{85} = 9.2\ldots$
so the triangle is isosceles.

5 Tony's ($\sqrt{4.8^2 + 5.3^2} = 7.2$)

6 Yes probably, $\sqrt{81^2 + 205^2} = 220$ cm

7 $4 \times \sqrt{3.5^2 + 6^2} = 4 \times 6.946\ldots = 27.8$ cm

8 (a) 7.1 cm   (b) 6.8 cm

⑭ Simultaneous equations

The emphasis here is on finding ways to compare two equations and solve them by looking at the **difference** between them. For example, to solve $3a + 2b = 24$ and $3a - b = 15$, first note that you can obtain the left-hand side of the first equation by adding $3b$ to the left-hand side of the second equation. Then note that you can obtain the right-hand side of the first equation by adding 9 to the right-hand side of the second equation. Hence $3b = 9$ and the solution can be found. More formal methods of adding and subtracting equations are not considered here and will be met later in the course.

p 95	**A** What's the difference?	Problems leading to solving simultaneous equations
p 96	**B** Addition and subtraction crosses	Introducing solving simultaneous equations by looking at the difference between the equations
p 99	**C** Spot the difference	Problems where the difference between the equations is harder to find
p 100	**D** Shall I compare thee...?	Problems where you need to multiply one or both equations by a constant
p 102	**E** Useful arrangements	More difficult problems where some rearrangement or substitution is needed

---

Practice booklet pages 42 to 44

Ⓐ **What's the difference?** (p 95)

◊ Pupils should solve these by noticing the difference between the situations in each case.

◊ The problems can then be discussed as part of the introduction to section B. Equations could be formed and solved for each problem.

Ⓑ **Addition and subtraction crosses** (p 96)

◊ Pupils could work on the four crosses in groups and the whole class can then discuss methods used. The discussion should lead to the following method for the last cross.

Choose letters (for example, a and b) for the numbers in the pink square (the 'pink number') and in the centre. The 'green number' can be labelled $4a$ and equations can be formed: $a + b = 13$ and $4a + b = 37$. Solve by noting that you can obtain the left- and right-hand sides of the second equation by adding $3a$ and 24 respectively to the left- and right-hand sides of the first equation. So $3a = 24$, $a = 8$ and substitution gives $b = 5$. Check by completing the whole cross.

Some pupils may suggest that if b is the middle number in the cross, the expression in the green square could be $37 - b$ but this would lead unhelpfully to $37 - b + b = 37$. This is an opportunity to discuss the difference between an identity and an equation.

Others may note that you can solve the problem by finding two different expressions for the number in the centre and equating them to obtain a linear equation with one variable. For example, if a is the number in the pink square then the centre number is equivalent to $13 - a$ or $37 - 4a$. So $13 - a = 37 - 4a$ leading to $a = 8$. These pupils could try both methods for the problems in B2 and decide which they prefer.

◊ In your introduction, include crosses that lead to equations such as $a + b = 4$ and $3a + b = 2$. Pupils need to notice that you can obtain the left hand side of the second equation by adding $2a$ to the left-hand side of the first equation and you can obtain the right-hand side of the second equation by adding $^-2$ to the right-hand side of the first equation, giving $2a = {}^-2$ (not $2a = 2$).

◊ Pairs of simultaneous equations can be formed and solved for the problems in section A.

*B11 (b) (i) Letting x be the green number and y the centre number leads to the equations $xy = 4$ and $(x + 4)y = 6$, or $xy + 4y = 6$.
From these it follows that $4y = 2$.

ℂ **Spot the difference** (p 99)

The approach described below is not the usual one of 'adding'.

◊ Ask pupils to form equations where p is Patsy's number and q is Quentin's ($3p + q = 40$ and $3p - q = 32$). Ask pupils what you need to add to the left-hand side of the second equation to obtain the left-hand side of the first. Hence obtain $2q = 8$ giving $q = 4$ and $p = 12$. Some pupils may see that you can 'add' the equations to obtain $6p = 72$ and high attainers could be encouraged to look at a variety of methods. However, many pupils will benefit from the consistent approach of looking at differences at this stage.

In your introduction, make sure that you include an example like C2 where both equations involve subtraction.

'I liked this very much. The topic went much more smoothly than in the previous year when I had a more able group.'

Secret numbers

Ask pupils each to think of two numbers, p and q. Then tell them to work out the values of, for example, $2p + 3q$ and $2p + 5q$.

Ask someone for their answers. The rest of the class has to work out the values of p and q.

D Shall I compare thee...? (p 100)

◊ Trying to find the prices of the different individual items leads to pairs of equations that cannot be directly compared. For example, the first pair of pictures leads to $3c + b = 145$ and $5c + 3b = 275$ (working in pence). By the end of the discussion, pupils should understand that you can multiply both sides of an equation by the same number and still have a true statement about the unknowns. The first equation can be multiplied by 3 to obtain $9c + 3b = 435$ (thinking of the price of 9 coffees and 3 buns might help pupils see this) and now we can compare this with $5c + 3b = 275$ to give $4c = 160$ and $c = 40$.

For the second pair of pictures, both equations need to be multiplied. Show two different ways of approaching this, for example multiplying by 2 and 3, or 5 and 2.

◊ Equations involving subtraction are covered from D8 and you may wish to include these in your initial introduction.

◊ Some pupils may spot that some equations can be simplified by dividing through by a common factor, for example, in question D4(f).

E Useful arrangements (p 102)

◊ The approach expected here is to ensure that both equations are in a form such that each side can be usefully compared. For example, in problem A we can add b to both sides of the second equation to give $a + b = 30$ and then work from there. We could of course subtract b from both sides of the first equation but this is likely to look more daunting.

Some pupils may see that you can substitute to obtain $6(30 - b) + b = 90$ and be quite happy to use this approach. However, many will find this confusing at this stage and it will be covered later in the course.

Ⓐ What's the difference? (p 95)

A1 (a) 120 grams   (b) 10 grams

A2 (a) 5 grams   (b) 160 grams

A3 A tea costs 80p and a bun costs 30p.

A4 (a) ♥ = 10   ■ = 2   ◆ = 1

    (b) The pupil's puzzles

Ⓑ Addition and subtraction crosses (p 96)

B1 (a) $3a$ (b) $a + b = 21$, $3a + b = 41$

    (c) $a = 10$, $b = 11$ and the whole cross

B2

	Pink	Green	Centre
(a)	7	28	3
(b)	40	8	25
(c)	14	7	$^-2$
(d)	$^-1$	$^-10$	11

B3 (a) $3a = \mathbf{6}$ (b) 2   (c) 5

    (d) $2 \times 2 + 2 \times 5 = 14$

B4 (a) $a = 4, b = 1$ (b) $a = ^-1, b = 9$

    (c) $a = 2, b = 2$ (d) $a = 3, b = 3$

    (e) $a = 5, b = ^-1$ (f) $a = 7, b = ^-2$

B5 (a) $4a = \mathbf{12}$ (b) 3   (c) 1

    (d) $5 \times 3 - 1 = 14$

B6

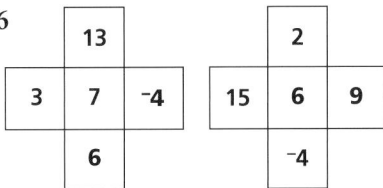

B7 (a) $4x$ (b) $x - y = 2$, $4x - y = 17$

    (c) $x = 5, y = 3$ and the whole cross

B8

	Pink	Green	Centre
(a)	6	60	4
(b)	62.5	12.5	9.5
(c)	2	8	$^-3$
(d)	1.5	0.5	0.5

B9 (a) $7a = \mathbf{28}$ (b) 4   (c) 1

    (d) $9 \times 4 - 3 \times 1 = 33$

B10 (a) $p = 4, q = 2$ (b) $p = 7, q = 3$

    (c) $p = 5, q = 1$ (d) $p = 2, q = 6$

    (e) $p = 8, q = 3$ (f) $p = 5, q = ^-2$

*B11 (a) 56

(b)

	Pink	Green	Centre
(i)	12	8	0.5
(ii)	$^-9$	$^-2$	$^-4$

Ⓒ Spot the difference (p 99)

C1 (a) $7y$ (b) $x = 2, y = 2$

C2 (a) $2y$ (b) $x = 10, y = 3$

C3 (a) $p = 5, q = 2$ (b) $a = 4, b = 3$

    (c) $c = 3, d = 3$ (d) $x = 6, y = 1$

    (e) $h = 5, k = 3$ (f) $m = 10, n = ^-3$

C4 (a) $d + n = 17$ (b) $n - d = 5$

    (c) $d = 6$ and $n = 11$ so David is 6 and Nicola is 11.

C5 $a - 2b = 9$ and $a + 3b = 24$; $a = 15, b = 3$

C6 Cindy is 8 years old.

Ⓓ Shall I compare thee…? (p 100)

D1 (a) $3a + 15b = 141$ (b) $a = 7, b = 8$

D2 (a) $6m + 4n = 30$ (b) $27n + 6m = 168$

    (c) $m = 1, n = 6$

D3 (a) (i) $8x + 20y = 56$ (ii) $25x + 20y = 90$

       (iii) $x = 2, y = 2$

    (b) $10x + 25y = 70$, $10x + 8y = 36$ leading to $x = 2, y = 2$

D4 (a) $v = 5, w = 2$ (b) $p = 1, q = 9$

    (c) $m = 3, n = 6$ (d) $k = 5, h = ^-1$

    (e) $x = 4, y = 4$ (f) $a = 3, b = ^-2$

D5 $2i + c = 90$, $3i + 5c = 205$ giving 20p as the cost of a candy bar

D6 An egg is 1 kg and a goose is 3 kg.

D7 (a) $10x - 25y = 30$ (b) $10x + 6y = 92$

    (c) $x = 8, y = 2$

D8 (a) $20a - 12b = 52$ (b) $21a - 12b = 57$
(c) $a = 5, b = 4$

D9 (a) $p = 5, q = 2$ (b) $a = 7, b = 3$
(c) $c = 6, d = {}^-1$ (d) $x = 4, y = 5$
(e) $h = 3.5, k = 0.5$ (f) $m = 2, n = {}^-3$

D10 $2s + t = 10, 3t - s = 9$ giving Simonetta's number as 3 and Tomaso's as 4

E Useful arrangements (p 102)

E1 (a) $x = 1, y = 4$ (b) $m = 5, n = {}^-2$
(c) $a = 19, b = 6$ (d) $p = 7, q = 3$
(e) $y = {}^-1, z = 2.4$ (f) $s = 3, t = 10$

E2 (a) $x = 2, y = 15$ (b) $p = 6, q = {}^-1$
(c) $m = 4, n = 3$ (d) $x = 5, y = 5$
(e) $g = 5, h = {}^-3$ (f) $a = 14, b = 6$

E3 Golden: 13.5 cents Green: 4.5 cents

What progress have you made? (p 102)

1 (a) $a = 5, b = 2$ (b) $x = 2, y = 10$
(c) $m = 3, n = 4$ (d) $p = 6, q = {}^-1$
(e) $v = 3, w = {}^-3$ (f) $h = 1.5, k = 0.5$

2 Dola: 20 grams Kwatro: 15 grams

Practice booklet

Sections B and C (p 42)

1 (a) $3y + 3b = 21,\ 3y + 5b = 29$
(b) $y = 3,\ b = 4$, so the yellow stick is 3 cm long and the brown 4 cm.

2 (a) $x = 4, y = 1$ (b) $x = {}^-1, y = 5$
(c) $x = {}^-1, y = 2$

3 (a) $x = 4, y = 3$ (b) $x = 2, y = {}^-1$
(c) $x = {}^-2, y = {}^-2$

4 $5g - b = 10,\ 5g + 3b = 30,\ g = 3,\ b = 5$
The green stick is 3 cm long and the black 5 cm.

5 (a) $x = 4, y = 1$ (b) $x = 6, y = 2$
(c) $x = 3, y = {}^-2$

6 (a) $l - b = 1,\ l + 3b = 17$
(b) $l = 5,\ b = 4$, so the lilac stick is 5 cm long and the brown 4 cm.

7 (a) $2m - p = 2,\ 2m + 3p = 18$
(b) $m = 3,\ p = 4$, so the mauve stick is 3 cm long and the pink 4 cm.

Section D (p 43)

1 (a) $3r + 2p = 24,\ r + 3p = 15$
(b) $r = 6,\ p = 3$, so the red brick weighs 6 g and the pink 3 g

2 (a) $5g + 2y = 9,\ 4g + 7y = 18$
(b) 36 g (c) 90 g
(d) $g = 1,\ y = 2$, so the green brick weighs 1 g and the yellow 2 g.

3 (a) $x = 2, y = 4$ (b) $x = 4, y = 4$
(c) $x = 1, y = 6$

4 (a) $x = 5, y = 1$ (b) $x = {}^-2, y = 7$
(c) $x = 6, y = {}^-2$

5 $5l - 4b = 14,\ 2l + 5b = 32$
$l = 6,\ b = 4$, so the lilac brick weighs 6 g and the brown 4 g.

6 $4b - 3m = 29,\ 2b - 4m = 12$
$b = 8,\ m = 1$, so the blue brick weighs 8 g and the mauve 1 g.

7 (a) $x = 6, y = 9$ (b) $x = 6, y = 1$
(c) $x = 7, y = {}^-1$ (d) $x = 4, y = {}^-1$
(e) $x = {}^-2, y = {}^-3$ (f) $x = 3, y = {}^-4$

Section E (p 44)

1 (a) $c = 4, d = 1$ (b) $g = {}^-1, h = 6$
(c) $p = 3, q = 4$ (d) $u = 7, v = 2$
(e) $w = 4, z = 1$ (f) $e = 15, f = 10$
(g) $b = 1, a = 1$ (h) $r = 6, s = 3$
(i) $k = {}^-1, l = 7$ (j) $k = {}^-1, l = 7$
(k) $i = 3, j = 2$ (l) $m = 4, n = {}^-1$

2 A gold brick weighs 2 g and a violet 8 g.

Review 2 (p 103)

1 1.38

2 (a) $\frac{9}{25}$ (b) $\frac{4}{25}$ (c) $\frac{12}{25}$

3 (a)

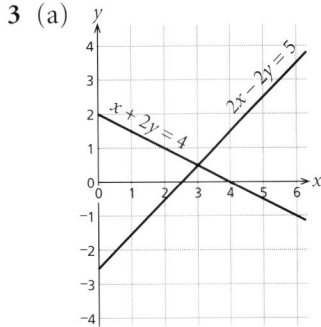

(b) $x = 3$, $y = \frac{1}{2}$

4 There should be a bigger gap between the 1996 and 2002 bars. The rate of growth is much less than between 1994 and 1996 but the chart makes it appear the same.

5 (a) $a = 6.7\,\text{cm}$ (b) $b = 4.1\,\text{cm}$
   (c) $c = 8.4\,\text{cm}$

6 (a) $x = 2.5$, $y = 1.5$
   (b) $x = {}^-1$, $y = 5$

7 (a)

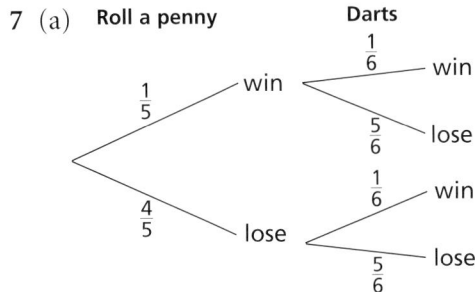

(b) (i) $\frac{1}{30}$ (ii) $\frac{9}{30}$ or $\frac{3}{10}$

8 32 minutes

9 $x = 7\frac{1}{2}$

10 (a) $p = 9.1$ (b) $49.28\,\text{m}^2$

11 (a) 0.26   (b) 0.85
    (c) 0.375   (d) 0.3125

12 Subtract 570 from each number. Work out the mean of 5, 9, 3, 7, 4, 8, which is 6.
    Add 570 to get **576**.

13 (a) $7.2\,\text{cm}$ (b) $162.6\,\text{cm}^2$

14 (a) 1 minute
    (b) $3\,\text{km/min}$
    (c) $180\,\text{km/h}$
    (d) (i) $80\,\text{km/h}$ (ii) $45\,\text{km/h}$
    (e) 1019

15 (a) $x = 1.25$ or $\frac{5}{4}$ (b) $x = {}^-1$
    (c) $x = 11$ (d) $x = 14$
    (e) $x = {}^-3$ (f) $x = {}^-8$

16 (a) Before: £1.60, after: £2.00
    (b) 25%

*17 Solve $8 - x = \frac{1}{2}(5 + x)$ to get $x = \frac{11}{3} = 3\frac{2}{3}$
     Graham gave $3\frac{2}{3}$ litres to Jane.

Mixed questions 2 (Practice booklet p 45)

1 1.6

2 (a) $\frac{2}{5}$ (b) $\frac{2}{15}$ (c) $\frac{7}{15}$

3 (a), (b)

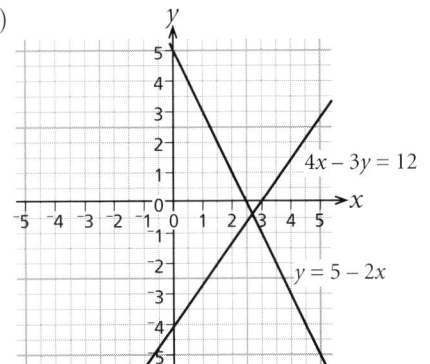

(c) $x = 2.7$, $y = {}^-0.4$ (d) $y = {}^-2 - 2x$

4 (a) $10.8\,\text{cm}$ (b) $7.2\,\text{cm}$
   (c) $35.2\,\text{cm}$ (d) $66\,\text{cm}^2$

5 (a) $x = 3$, $y = {}^-1$ (b) $a = 2$, $b = 3$

(c) $s = 4$, $t = 5$

6 Large: 24 balloons, medium: 15 balloons

7 71.9 kg

8 (a) Reflection in line $x = 3$

(b) Rotation of 180° with centre (4, 3)

(c) Translation $\begin{bmatrix} 5 \\ {}^-4 \end{bmatrix}$

(d) Rotation of 90° anticlockwise with centre (1, 3)

9 (a)

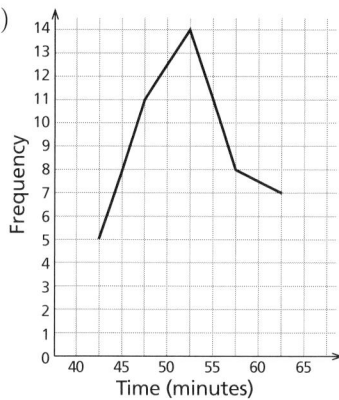

(b) Approximately 2367.5 ÷ 45
= 52.6, or 53 minutes

⑮ Spot the errors (p 106)

This is intended to help pupils to look critically at their own work and thus avoid errors.

1 The order of operations is wrong.
$78 \div (23 \times 1.9) = 1.78$ (to 2 d.p.)

2 18 tables are needed.

3 The hypotenuse c must be greater than each of the other two sides.
$c = 8.60$ cm (to 2 d.p.)

4 $5.95\ldots$ hours = 5 hours 57 minutes (to the nearest minute)

5 4 hours 30 minutes = 4.5 hours. The (incorrect) result is given to an unnecessary (and unjustified) degree of accuracy. $210 \div 4.5 = 46.7$ m.p.h. is sufficient.

6 The amount in $ has to be divided by 1.63: $5.70 \div 1.63 = £3.50$ (to the nearest penny).

7 The intermediate results have been rounded, giving rise to inaccuracy.
$107.5 \div 15.48 = 6.94$ cm (to 2 d.p.) or 6.9 cm (to 1 d.p.)

8 There are two errors.
0.2×0.3 should be 0.06;
$1 \text{ m}^2 = 10\,000 \text{ cm}^2$.
The area is $0.06 \times 10\,000 = 600 \text{ cm}^2$.

*9 The overall average speed is the total distance divided by the total time.
$84 \div 9.5 = 8.84$ m.p.h. (to 2 d.p.)

16 Changing the subject

The approach in this unit uses flow diagrams. This method works only when the new subject of the formula appears once in the original formula. However, the practice in the use of inverses will be useful at the later stage when the approach is widened.

Practice booklet pages 47 to 48

A From formula to equation (p 108)

If a formula is to be used 'in reverse' only once, it is simpler to substitute for the known values and solve the resulting equation.

B Reversing the flow (p 109)

In this section and the next, only two letters appear in each formula.

B9 If pupils get stuck here, the next section will help.

C Squares and square roots (p 110)

◊ The fact that $x^2 = a$ ($a > 0$) has two solutions, $x = \pm\sqrt{a}$, needs emphasis, although in many instances the negative solution will not fit the original problem.

D Formulas with several letters (p 110)

◊ This may be where some pupils lose their way. The first step is to identify the new subject in the formula; then the flow diagram has to show what is done to that variable, in the correct order. Lastly the flow diagram is reversed to give the rearranged formula.

You may find it helps to do two worked examples in parallel, one with numbers and one with letters (for example, $q = 3p - 7$ alongside $q = ap - b$).

Ⓐ From formula to equation (p 108)

A1 (a) $N = 60$ (b) $h = 3$
    (c) (i) $h = 9$ (ii) $h = 17$

A2 (a) $h = 3$ (b) $h = 12$ (c) $h = 14$

A3 (a) $N = 6h + 3$
    (b) (i) $h = 8$ (ii) $h = 13$ (iii) $h = 18$
    (c) 21

A4 (a) $N = 11l + 6$
    (b) (i) $l = 9$ (ii) $l = 14$

Ⓑ Reversing the flow (p 109)

B1 (a) $h \longrightarrow \boxed{\times 13} \longrightarrow \boxed{+ 7} \longrightarrow N$

    (b) $h = \dfrac{N - 7}{13}$

B2 (a) $p \longrightarrow \boxed{- 5} \longrightarrow \boxed{\times 2} \longrightarrow s$

    (b) $p = \dfrac{s}{2} + 5$

B3 (a) $v = \dfrac{t}{3} - 4$ (b) $t = 3(v + 4)$

B4 $r = 6m + 7$

B5 $a = 2(b - 9)$

B6 (a) $x = \dfrac{y + 2}{3}$ (b) $x = \dfrac{y}{3} + 2$
    (c) $x = 5y - 4$ (d) $x = 3(y + 8)$

B7 (a) $x \longrightarrow \boxed{- 2} \longrightarrow \boxed{\times 3} \longrightarrow \boxed{\div 5} \longrightarrow y$

    (b) $x = \dfrac{5y}{3} + 2$

B8 (a) $x \longrightarrow \boxed{\times 2} \longrightarrow \boxed{+ 5} \longrightarrow \boxed{\div 4} \longrightarrow y$

    (b) $x = \dfrac{4y - 5}{2}$

B9 $x = \pm \sqrt{y - 5}$

Ⓒ Squares and square roots (p 110)

C1 (a) $l \longrightarrow \boxed{\text{square}} \longrightarrow \boxed{\times 6} \longrightarrow S$

    (b) $l = \sqrt{\dfrac{S}{6}}$

C2 (a) $u = \pm \sqrt{v - 4}$ (b) $u = \pm \sqrt{v} + 2$

    (c) $u = \pm \sqrt{3v - 7}$ (d) $u = \pm \sqrt{\dfrac{v - 3}{5}}$

    (e) $u = (v + 3)^2$ (f) $u = \dfrac{v^2 + 1}{2}$

    (g) $u = \pm \sqrt{(v - 2)} - 5$

    (h) $u = \pm \sqrt{\dfrac{v}{2}} + 3$

Ⓓ Formulas with several letters (p 110)

D1 (a) $l \longrightarrow \boxed{+ w} \longrightarrow \boxed{\times 2} \longrightarrow P$

    (b) $l = \dfrac{P}{2} - w$

    (c) $l = 4$

D2 (a) $x = \dfrac{y - b}{a}$ (b) $p = \dfrac{q}{k} - 4$

    (c) $u = \dfrac{s}{a} + b$ (d) $x = d(y + b)$

    (e) $u = bw - a$ (f) $t = r\left(\dfrac{z}{a} - p\right)$

    (g) $x = \dfrac{cy + b}{a}$ (h) $p = \dfrac{mq}{k} - h$

D3 (a) $a = \sqrt{\dfrac{V}{l}}$ (b) $l = \dfrac{S - 2a^2}{4a}$

D4 $V = \pm \sqrt{PR}$

*D5 $r = \sqrt[3]{\dfrac{3V}{4\pi}}$

What progress have you made? (p 111)

1 (a) $x = 4(y - 5)$ (b) $x = \dfrac{3y + 2}{5}$

2 (a) $p = qs + r$ (b) $p = b\left(\dfrac{q}{a} - c\right)$

3 $r = \sqrt{\dfrac{S}{4\pi}}$

Practice booklet

Sections A and B (p 47)

1 (a) 40

  (b) (i) 24   (ii) 54

2 (a) $J = 3h + 3$

  (b) (i) 11   (ii) 33

  (c) 52

3 (a) $h \longrightarrow \boxed{\times 9} \longrightarrow \boxed{+ 3} \longrightarrow J$

  (b) $h = \dfrac{J - 3}{9}$

4 (a) $p = \dfrac{C + 5}{2}$ (b) $C = 2p - 5$

5 (a) $r = \dfrac{t - 6}{5}$ (b) $r = \dfrac{t}{4} - 1$

  (c) $r = 3t + 9$ (d) $r = 5(t - 3)$

6 (a) (i) $x \longrightarrow \boxed{\times 5} \longrightarrow \boxed{- 4} \longrightarrow \boxed{\div 2} \longrightarrow y$

    (ii) $x = \dfrac{2y + 4}{5}$

  (b) (i) $x \longrightarrow \boxed{+ 2} \longrightarrow \boxed{\times 6} \longrightarrow \boxed{\div 7} \longrightarrow y$

    (ii) $x = \dfrac{7y}{6} - 2$

Sections C and D (p 48)

1 (a) $t \longrightarrow \boxed{\text{square}} \longrightarrow \boxed{\times 5} \longrightarrow s$

  (b) $t = \sqrt{\dfrac{s}{5}}$

2 (a) $c = \pm\sqrt{d + 5}$ (b) $c = \pm\sqrt{d} - 6$

  (c) $c = \pm\sqrt{\dfrac{d - 9}{2}}$ (d) $c = \dfrac{\pm\sqrt{d} - 1}{2}$

  (e) $c = \pm\sqrt{5(d + 2)}$ (f) $c = \pm\sqrt{\dfrac{d}{5}} - 1$

3 (a) $x = \dfrac{p}{3} - q$ (b) $x = \dfrac{v - 5}{a}$

  (c) $x = \dfrac{4d + u}{3}$

4 (a) $x = \dfrac{y + q}{p}$ (b) $x = \dfrac{t}{r} - s$

  (c) $x = ac - b$ (d) $x = r(p - q)$

  (e) $x = \dfrac{bv}{a} + u$ (f) $x = \dfrac{cg + h}{a}$

5 (a) $h = \dfrac{3V}{x^2}$ (b) $x = \sqrt{\dfrac{3V}{h}}$

⑰ Similar shapes

> **Essential**
>
> A2 paper, broadsheet newspaper pages, metre rules
> Sheet 251, cotton, small weight
> Tape measure or trundle wheel
>
> **Practice booklet** pages 49 to 52

Ⓐ Scaling (p 112)

◊ Not all the dimensions have been doubled. Also, the angles of the rhombus are different in the copy – if the scaling had been done correctly the angles would be the same as in the original.

◊ In the second copy, the width has been enlarged from 5 cm to 12.5 cm, a scale factor of 2.5. So the height of this copy is 4 cm × 2.5 = **10 cm**.

Ⓑ Scaling down (p 114)

Ⓒ Ratios within shapes (p 116)

> A2 paper, broadsheet newpaper pages, metre rule

Paper sizes

When one of the A series of paper sizes is folded in half by a fold parallel to the shorter edge, each half is mathematically similar to the original. If the sheets are laid out as shown on the next page, it can be seen that each is a scaled copy of the others.

The ratio of the longer to the shorter side is the same ($\sqrt{2}$) for all the sizes.

In the case of the broadsheet size, the 'half-sheet' (tabloid) is not a scaled copy of the original and the ratio of longer to shorter side is not the same.

However, when the tabloid size is halved, the resulting shape *is* a scaled copy of the broadsheet. (The same is true for any starting shape – the 'quarter-size' will be similar to the original.)

(The A series is such that size A0 has an area of $1 \, m^2$.)

An advantage of the A series is that any size can be enlarged or reduced to fit another.

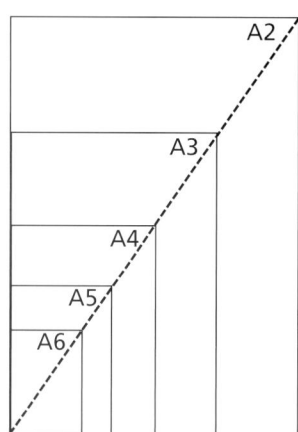

◊ Pupils need to understand that the ratio $\frac{B}{A}$ is the multiplier from A to B.

D Scale factors and ratios (p 118)

◊ Emphasise the distinction between the two types of comparison:
  • comparison of copy to original (scale factor)
  • comparison of two selected lengths within the original and the copy (equal ratios)

Higher attainers should be able to appreciate an algebraic explanation of the connection between the two:
  • Suppose the scale factor is k.
    Then two lengths p, q in the original become kp, kq in the copy.
    The ratio $\frac{p}{q}$ in the original becomes $\frac{kp}{kq}$ which is equal to $\frac{p}{q}$.

E Similarity (p 119)

◊ When the corresponding angles of two triangles are equal, the triangles are similar. The same is not true for quadrilaterals or other polygons. For example, these two trapezia are equiangular but not similar:

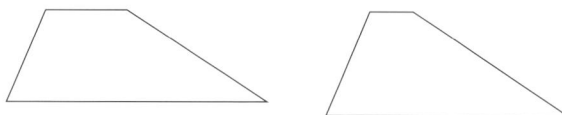

Finding the heights of tall objects

Sheet 251, cotton, small weight, tape measure or trundle wheel

◊ The ratiometer on sheet 251 uses the idea of similar triangles to find the heights of tall objects. This is, of course, a lead-in to the idea of the tangent of an angle.

Ⓐ Scaling (p 112)

A1 (a)

Measurement	Original length	× ?	Length in copy
Height of building	**4.2 cm**	**1.5**	**6.3 cm**
Length of ladder	**3.2 cm**	**1.5**	**4.8 cm**
Height of door	**2.0 cm**	**1.5**	**3.0 cm**
Width of door	**1.0 cm**	**1.5**	**1.5 cm**

(b) 1.5

(c) Both 72°

A2 (a) Base 3 cm, height 2 cm

(b) A: base 6 cm,   height 4 cm
B: base 6.5 cm,   height 5 cm
C: base 3.6 cm,   height 2 cm
D: base 4.5 cm,   height 3 cm

(c) (i) A: 2, B: 2.16, C: 1.2, D: 1.5

(ii) A: 2, B: 2.5, C: 1, D: 1.5

(d) A and D

A3

	Original length	Scale factor	Copy length
Width of picture	10 cm	**2.5**	25 cm
Height of picture	6 cm	**2.5**	**15 cm**
Height of house	3 cm	**2.5**	**7.5 cm**
Length of car	**2 cm**	**2.5**	5 cm
Length of hedge	**8 cm**	**2.5**	20 cm

A4 224 mm

A5 (a) A: 1.25 B: 1.75 C: 2.4

(b)

Measurement	Original length	Length in copy A	Length in copy B	Length in copy C
Karl's height	20 mm	25 mm	**35 mm**	**48 mm**
Height of house	120 mm	**150 mm**	210 mm	**288 mm**
Height of tree	150 mm	**187.5 mm**	**262.5 mm**	360 mm

Ⓑ Scaling down (p 114)

B1 (a) $\frac{1}{2}$ (b) $\frac{1}{6}$

(c) $\frac{1}{3}$ (d) $\frac{2}{3}$

B2 (a) 0.45   (b) 0.65

(c) 0.36   (d) 0.83

B3

	Original length	Scale factor	Copy length
Width of picture	8 cm	**0.6**	4.8 cm
Height of picture	5 cm	**0.6**	**3 cm**
Height of tree	3 cm	**0.6**	**1.8 cm**
Length of pond	**7.5 cm**	**0.6**	4.5 cm
Length of fence	**9 cm**	**0.6**	5.4 cm

B4 (a) A: 0.8 B: 0.4 C: 0.55

(b)

Measurement	Original length	Length in copy A	Length in copy B	Length in copy C
Angela's height	25 mm	20 mm	**10 mm**	**13.75 mm**
Length of flute	10 mm	**8 mm**	4 mm	**5.5 mm**
Height of music stand	20 mm	**16 mm**	**8 mm**	11 mm

Ⓒ Ratios within shapes (p 116)

C1 (a) AB = 6 cm   DE = 9 cm
Scale factor 1.5

(b) BC = 3.6 cm   EF = 5.4 cm
Scale factor $\frac{2}{3}$

(c) Scale factor 1.5

(d) ABC = 60°; DEF is also 60°.

(e) BCA = 83°; EFD is also 83°.

(f) CAB and FDE are both 37°.

C2 A and C have ratio 1.6
B and G have ratio 1.33
D and H have ratio 1.25
E and F have ratio 1.88

C3

Height of picture	Ratio $\frac{width}{height}$	Width of picture
5 cm	**1.6**	8 cm
7.2 cm	**1.6**	**11.52 cm**
7.5 cm	**1.6**	12 cm

C4

Height of picture	× ?	Width of picture
18 cm	**0.67**	12 cm
8.4 cm	**0.67**	**5.6 cm**
30 cm	**0.67**	20 cm

D Scale factors and ratios (p 118)

D1 (a) $\frac{50}{25} = \frac{21}{10.5} = 2$

(b) $\frac{25}{10.5} = \frac{50}{21} = 2.38$

D2 (a) 0.525

(b) (i) 75 mm (ii) 26.25 mm

(iii) 62.5 mm

E Similarity (p 119)

E1 P, Q, R and S all have ratio 1.6.

E2 The angles are the same.

E3 A and D have ratio 0.75.

E4 (a) 2.4

(b) A: 24 cm   B: 14.4 cm   C: 20.4 cm

E5 (a) 0.6

(b) $p = 3.6$ cm, $q = 2.7$ cm, $r = 3.9$ cm,
$s = 7$ cm, $t = 3.\dot{3}$ cm

What progress have you made? (p 122)

1 A: 1.4   B: 0.6

2 (a) 0.8

(b) $r = 5.2$ cm

(c) $s = 4.8$ cm

Practice booklet

Sections A and B (p 49)

1 2

2

	Rammi's plan	× ?	Alex's plan
Height of box (h)	9.6 cm	**1.5**	14.4 cm
Height of box (w)	6.8 cm	**1.5**	**10.2 cm**
Diameter of hole (d)	**3.2 cm**	**1.5**	4.8 cm
Height of roof (r)	4.2 cm	**1.5**	**6.3 cm**

3 (a) C and D

(b) A to C   × 0.4
A to D   × 0.8

Sections C and D (p 50)

1 (a) A and F, and E and G,
B and D, C and H,

(b) The pupil's check

2 (a) 1.25

(b)

Height (h)	Width (w)
5.8 cm	**7.25 cm**
12.6 cm	**15.75 cm**
6.08 cm	7.6 cm

3

Height (h)	Width (w)	$\frac{w}{h}$
30 cm	24 cm	**0.8**
5.25 cm	4.2 cm	**0.8**
7 cm	**5.6 cm**	**0.8**

Section E (p 51)

1 (a) 1.6182

(b) $\frac{BC}{EB} = \frac{55}{34} = 1.6176$

So BCFE and ABCD are very nearly
similar.

2 $a = 3.6$ cm

$b = 6.3$ cm

$c = 3.6$ cm

⑱ Functions and graphs

Essential

Graph paper, graph drawing computer program or graphical calculator
Sheets 252 and 253

Practice booklet page 53

Ⓐ Quadratic functions (p 123)

Graph paper, sheet 252

Ⓑ Other functions (p 125)

Graph paper, sheet 253

B2 You can introduce the word 'asymptote' to describe a line that a graph gets closer and closer to without reaching.

The asymptotes of $y = \frac{a}{x}$ are the x- and y-axes.

Investigation

The most obvious effects are seen when c is changed from positive to negative and when a is changed from positive to negative.
Changing c to negative translates the graph $2c$ units downwards.
Changing a, b and c to negative reflects the graph in the x-axis.
Changing a to negative turns the graph upside down, but depending on the values of b and c, translates it as well.

A Quadratic functions (p 123)

A1 (a)

x	-3	-2.5	-2	-1.5	-1	-0.5	0	0.5	1	1.5	2	2.5	3
y	9	6.25	**4**	**2.25**	**1**	**0.25**	0	**0.25**	**1**	**2.25**	**4**	**6.25**	**9**

(b) The pupil's graph of $y = x^2$

(c) It is symmetrical about the y-axis.

(d) x is about 2.4.

(e) Find x when $y = 3$ (x is about 1.7).

A2 (a)

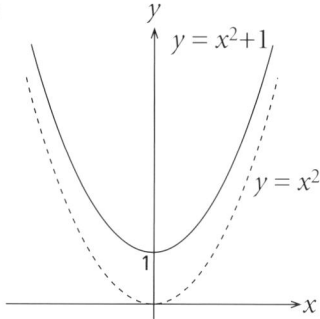

(b) The pupil's graph of $y = x^2 + 1$

A3 (a)

x	-3	-2	-1	0	1	2	3
y	7	2	-1	-2	-1	2	7

(b) The pupil's graph of $y = x^2 - 2$

(c) About $^-1.4$ and 1.4

A4 (a) $y = x^2 + 2$ (b) $y = x^2 - 1$

(c) $y = {}^-x^2$

A5

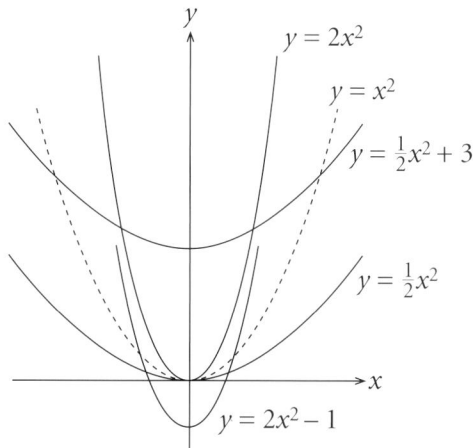

****A6** (a) $y = \frac{1}{4}x^2$ (b) $y = 2x^2 - 3$

B Other functions (p 125)

B1

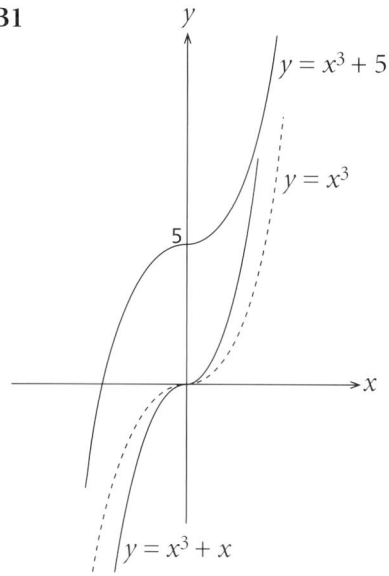

B2 (a)

x	-6	-5	-4	-3	-2	-1	1	2	3	4	5	6
y	-1	-1.2	-1.5	**-2**	**-3**	**-6**	**6**	**3**	**2**	**1.5**	**1.2**	**1**

(b) Points plotted

(c) (i) $y = 12$ (ii) $y = 60$ (iii) $y = 600$

(d) (i) $y = {}^-12$ (ii) $y = {}^-60$ (iii) $y = {}^-600$

(e) You cannot divide by 0.

(f)

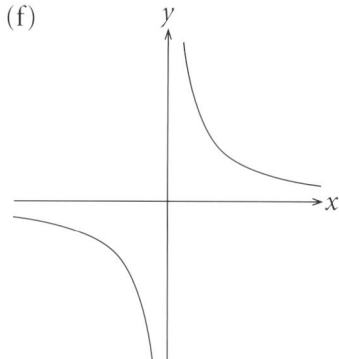

B3 (a) $(2, {}^-2\frac{1}{2})$ (b) $({}^-\frac{1}{2}, \mathbf{10})$

B4 The pupil's results from the computer program

What progress have you made? (p 126)

1 (a)

x	‾4	‾3	‾2	‾1	0	1	2	3	4
y	‾4	3	8	11	12	11	8	3	‾4

(b)

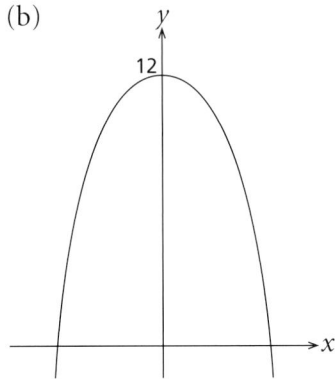

(c) About ‾3.5 or 3.5

2 (a)

x	‾3	‾2	‾1	0	1	2	3
y	‾30	‾11	‾4	‾3	‾2	5	24

(b), (c)

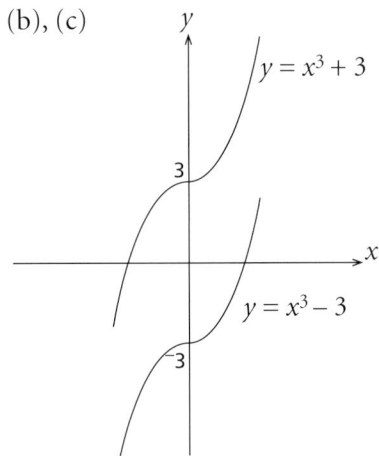

3 (a) $(3, \textbf{4})$ $(\textbf{‾4}, ‾3)$ $(‾3, \textbf{‾4})$ $(4, \textbf{3})$

(b)

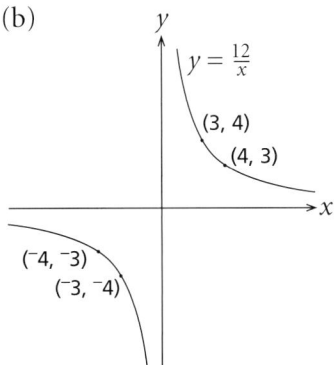

Practice booklet

Sections A and B (p 53)

1 (a)

x	‾3	‾2	‾1	0	1	2	3
$4 - x^2$	‾5	0	3	4	3	0	‾5

(b), (c)

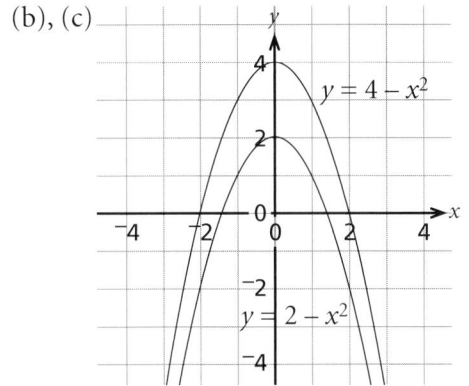

2 $y = x^2$ B
$y = x^2 + 3$ C
$y = 3x^2$ D
$y = \frac{1}{3}x^2$ A

3 (a)

x	‾3	‾2	‾1	‾0.5	0	0.5	1	2	3
x^3	‾27	‾8	‾1	‾0.125	0	0.125	1	8	27
$x^3 - 5$	‾32	‾13	‾6	‾5.125	‾5	‾4.875	‾4	3	22

(b)

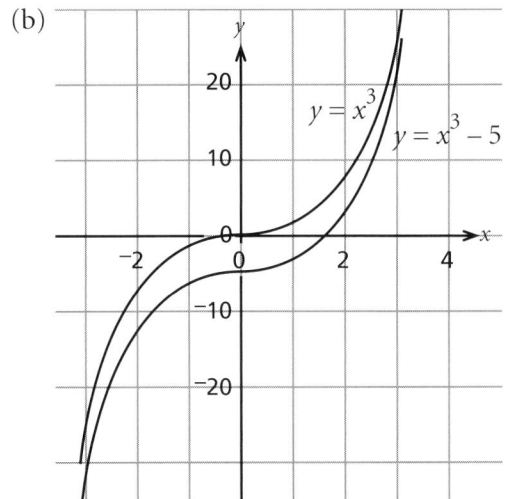

(c) 1.7

4 (a)

x	0.1	0.2	0.4	0.5	1	2	4	10
$\dfrac{2}{x}$	20	10	5	4	2	1	0.5	0.2

x	⁻10	⁻4	⁻2	⁻1	⁻0.5	⁻0.4	⁻0.2	⁻0.1
$\dfrac{2}{x}$	⁻0.2	⁻0.5	⁻1	⁻2	⁻4	⁻5	⁻10	⁻20

(b)

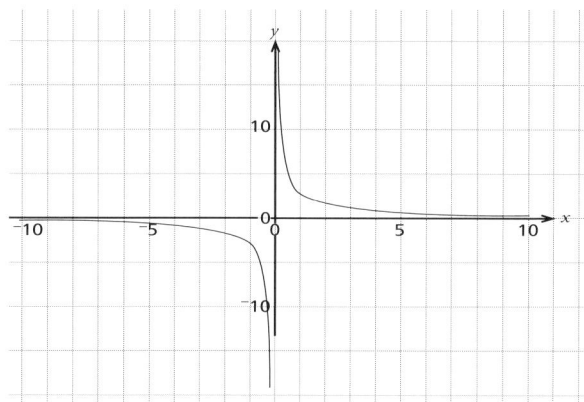

(c) Approximately ⁻0.3

⑲ Inequalities

Practice booklet pages 54 to 56

Ⓐ Simple inequalities (p 127)

◊ Explain the symbols $<$, $>$, $\leq$, $\geq$ and how to represent simple inequalities on a number line. Pupils can then try to match up as many of the statements and diagrams A to J as they can.

The correct matches are A with J, E with C, I with B and G, D with H.
F is unmatched.

A1 Some pupils may feel, for example in part (g), that $0.19 \leq 0.2$ is false because $0.19 \neq 0.2$. It might help to point out that $0.19 \leq 0.2$ means that **either** $0.19 < 0.2$ **or** $0.19 = 0.2$ and since $0.19 < 0.2$ is true then the whole statement $0.19 \leq 0.2$ is true.

Ⓑ Combined inequalities (p 128)

◊ Pupils can again try to match up as many of the statements and diagrams A to J as they can.

The correct matches are A with B and G, E with D and J, I with C.
H and F are unmatched.

Ⓒ In words and in symbols (p 130)

◊ Emphasise that you need to be precise in the definition of letters; for example, $r \geq 5$ does not mean red biros are greater than or equal to 5. This is too vague and could refer to the length or weight of red biros for example. In this case you need to state that r stands for the **number** of red biros in the bag.

C3 In part (f), you may need to point out that usually 'between m and n' includes m and n.

Ⅾ Challenges (p 132)

Ⓐ Simple inequalities (p 127)

A1 (a) T (b) F (c) T (d) F
(e) T (f) T (g) T (h) F
(i) F (j) T (k) T (l) T

A2 (a) $\sqrt{\frac{1}{4}}, \frac{2}{5}, 1, ^-20$
(b) $\sqrt{10}, \pi, \frac{13}{4}$

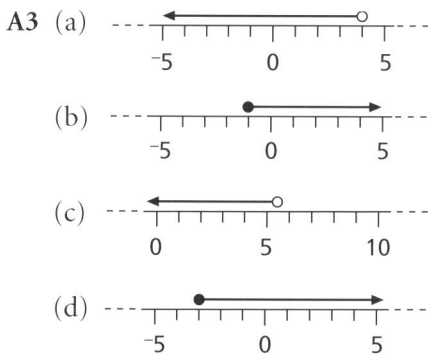

A3 (a)

-5 0 5

(b)

-5 0 5

(c)

0 5 10

(d)

-5 0 5

A4 (a) $x \geq ^-4$ (b) $x < 0$
(c) $x \leq 2.5$ (d) $x > ^-2$

A5 (a) Sometimes   (b) Always
(c) Always   (d) Never
(e) Always   (f) Sometimes
(g) Always   (h) Never

A6 (a) Never   (b) Sometimes
(c) Always   (d) Sometimes
(e) Always   (f) Never
(g) Always   (h) Sometimes

A7 The pupil's five numbers less than 3

A8 The pupil's four positive numbers less than or equal to 1.2

A9 97

A10 (a) $^-3, ^-2, ^-1, 0, 1, 2, 3$
(b) $^-4, ^-3, ^-2, ^-1, 0, 1, 2, 3, 4$
(c) $^-6, ^-5, ^-4, ^-3, ^-2, ^-1, 0, 1, 2, 3, 4, 5, 6$

Ⓑ Combined inequalities (p 128)

B1 (a) $5 < x < 10$ (b) Not possible
(c) $^-4 \leq x < 3$ (d) $0 \leq x < 9$
(e) $x < 1$ (f) $^-9 \leq x \leq 12$
(g) Not possible   (h) $x \geq 3$

B2 (a) $\sqrt{4}, \sqrt{24}$ (b) $^-2, 0, \sqrt{4}$

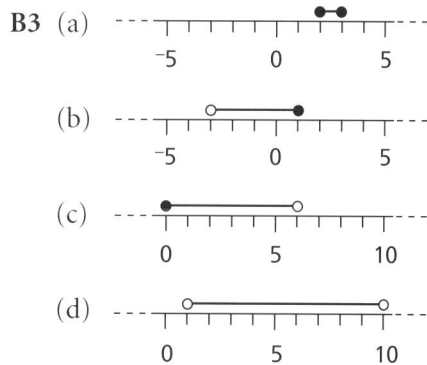

B3 (a)

-5 0 5

(b)

-5 0 5

(c)

0 5 10

(d)

0 5 10

B4 (a) $^-3 \leq x \leq 4$ (b) $^-5 < x \leq 0$
(c) $1 < x < 6$ (d) $^-5 \leq x < ^-1$

B5 (a) Sometimes   (b) Always
(c) Never   (d) Always

B6 The pupil's five numbers between $^-1$ and 2 (not including 2)

B7 The pupil's four numbers between 4 and 5

B8 The pupil's two numbers between 1 and 1.5 (not including 1.5)

B9 (a) 4, 5, 6, 7, 8   (b) 4, 5, 6, 7, 8

(c) $^-1$ (d) 2, 3

(e) 1   (f) 0

(g) $^-2, ^-1, 0$ (h) 7, 8, 9

(i) $^-3, ^-2, ^-1, 0, 1, 2, 3$

B10 (a) $^-6, ^-5, ^-4, ^-3, 3, 4, 5, 6$

(b) $^-7, ^-6, ^-5, 5, 6, 7$

ℂ **In words and in symbols** (p 130)

C1 (a) $s \geq 20$ or $20 \leq s$

(b) $m > 40$ or $40 < m$

(c) $s \leq 35\,000$ or $35\,000 \geq s$

(d) $s > c$ or $c < s$

(e) $r > w + p$ or $w + p < r$

C2 W matches with A.
p stands for the weight of popcorn in a bag.

X matches with D and I.
p stands for the number of passengers in the coach at any one time.

Y matches with G.
p stands for the prime number and m stands for the number Mark is thinking of.

Z matches with C and H.
p stands for the number of pins and m stands for the number of matches.

C3 (a) $5h > 3.60$ where h is the cost of a hot-dog in pounds

(b) $d < \frac{1}{2}c$ (or $2d < c$) where d is the number of dogs and c is the number of cats

(c) $d \geq 2s$ where d is the number of ducks and s is the number of swans

(d) $3c + 4d = 5b$ where c is the cost of a bar of chocolate, d is the cost of a can of drink and b is the cost of a burger

(e) $3c > 7l$ where c is the weight of a bottle of cola and l is the weight of a small bottle of lemonade

(f) $190 \leq c \leq 210$ where c is the weight of a chocolate bar

(g) $h > 1.5$ where h is your height in metres

(h) $d \geq c + 5$ (or $d - 5 \geq c$) where d is the number of dogs and c is the number of cats

(i) $12 \leq t < 17$ where t is the time in seconds John took to run 100 metres

C4 The pupil's statements

C5 (a) (i) £50   (ii) £$(6n + 20)$

(b) (i) $6n + 20 \leq 100$ (ii) 13 packs

C6 $2 \leq a < 12$ where a is the age in years

C7 $30 + 3.5t \leq 60$,   8 hours

𝔻 **Challenges** (p132)

*D1 4

*D2 7

*D3 14

*D4 15

*D5 4

*D6 19

*D7 $^-4$

*D8 1, 2, 3, 4, 5, 6, $^-2$, $^-3$, $^-4$, $^-5$

*D9 2, 3, 4, 5

*D10 4, 5, 6

*D11 15, 20, 25, 30, 35, 40, 45, 50

*D12 (a) 0   (b) 25   (c) 20   (d) $^-12$

*D13 Examples are $\frac{1}{2}, \frac{7}{15}, \frac{8}{15}$

*D14 2, 3, 4, 5, 6, 7, 8, 9, 10, 11

*D15 There is no single smallest value. For any value proposed that is greater than 6.1, a smaller value can be given that is still greater than 6.1.

*D16 (a) $30 < ab < 40$

(b) $50 < 2a + 2b < 70$ or $25 < a + b < 35$

(c) $5 < \sqrt{a^2 + b^2} < 10$ or
$25 < a^2 + b^2 < 100$

What progress have you made? (p 133)

1 (a) T   (b) F   (c) T

2 (a)

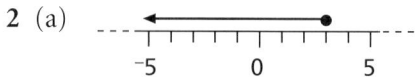

(b)

(c)

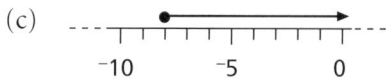

3 $x \leq 1$

4 14

5 (a) $4 \leq x < 10$ (b) $^-5 < n < 3$

6 (a)

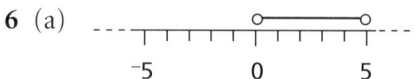

(b)

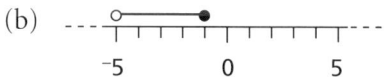

7 $^-3 \leq x < 5$

8 The pupil's four numbers between 1 and $3\frac{1}{3}$ (not including $3\frac{1}{3}$)

9 2, 3, 4, 5, 6

10 (a) $s \geq 40$ where s is the number of sweets in a bag

(b) $b > r$ where b is the number of black sweets and r is the number of red sweets

(c) $100 \leq w < 120$ where w is the weight of the bag of sweets in grams

(d) $g \geq 2r$ where g is the number of green sweets and r is the number of red sweets

Practice booklet

Section A (p 54)

1 (a) True   (b) False

(c) True   (d) True

(e) True   (f) True

(g) False   (h) True

2 (a)

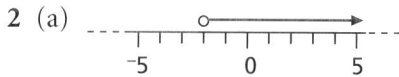

(b)

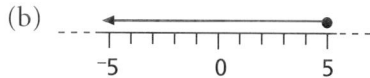

(c)

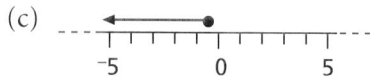

(d)

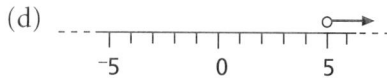

3 (a) $x \geq {}^-2$ (b) $x > 0$

(c) $x < 1$ (d) $x \leq 4$

4 The pupil's four values greater than 4

5 The pupil's three values less than or equal to 1

6 The pupil's three integers less than $1\frac{1}{3}$

7 28

8 53

9 4, 5, 6

10 17

11 $^-5$

12 72

13 3

14 2

15 10

Section B (p 55)

1 (a) $2 < x < 5$ (b) Not possible

(c) $^-1 < c \leq 7$ (d) Not possible

(e) $0 \leq s < 1$ (f) $^-7 < v < {}^-5$

2 $\sqrt{5}, 0, \frac{1}{2}\pi, \frac{4}{5}$

3 (a) $^-3 \le x \le 4$ (b) $^-4 \le x < 1$

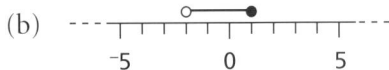

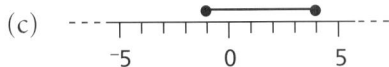

4 (a)

```
--- |--|--|--|--|--|--|--|--|--|--|--| ---
        -5        0    ○————○  5
```

     (b)

```
--- |--|--|--|--|--|--|--|--|--|--|--| ---
        -5     ○——●       5
               0
```

     (c)

```
--- |--|--|--|--|--|--|--|--|--|--|--| ---
        -5    ●————————●
              0         5
```

5 (a) 3, 4   (b) $^-1, 0, 1$

    (c) $^-4, ^-3$ (d) 4, 5

    (e) 3, 2   (f) 3

6 The pupil's three numbers between $\frac{1}{2}$ and 1

7 3, 4, 5

8 The pupil's three numbers greater than or equal to $^-0.5$ but less than 2

9 1, 2, 3, 4, 5, 6, 8, 10, 12, 15, 20

10 41, 43, 47, 53, 59

11 1, 2, 3, 4, $^-1, ^-2, ^-3, ^-4$

12 The pupil's fraction between $\frac{1}{4}$ and $\frac{1}{3}$.

13 3, 4, 5

*14 Maximum 0.3   Minimum $^-0.6$

Section C (p 56)

  1 (a) $d \ge 20$ where d is the number of drumsticks

      (b) $d \ge t + 20$ where d is the number of cards Dave has and t is the number of cards Tom has

      (c) $d + t < 20$ where d is number of cards Darina has and t is number of cards Tania has.

      (d) $d \ge 20t$ where d is the number of fruit drops and t is the number of tins (This is a situation that is particularly prone to confusion. To clarify it, ask if there will be more tins or more fruit drops.)

2 (a) $e \le 2$ where e is the number of errors

      (b) $p > 1500$ where p is the number of pupils in the school

      (c) $b > g$ where b is the number of boys and g is the number of girls

      (d) $8 \le a \le 13$ where a is the number of apples

      (e) $p \ge 2j$ where p is the number of CDs Paul has and j is the number of CDs that Jim has

      (f) $2l < 3s$ where l is the number of apples in the large pack and s is the number in the small pack

      (g) $a \le s + 3$ where a is the number of pages in Anita's essay and s is the number of pages in Sue's essay (or $s < a \le s + 3$)

      (h) $240 \le c < 350$ where c is the capacity of the can

3 The pupils' statements matching the inequalities

4 (a) $10 \le p \le 50$ where p is the number of pens

      (b) $2m + k \le 100$ where m is the number of mugs and k is the number of keyrings

20 More manipulation

Practice booklet pages 57 to 59

A Simplifying: review (p 134)

B Further simplification (p 135)

◊ The first diagram could be split by a vertical line to find the area. Pupils could check that this gives the same expression.

◊ Some pupils may try to 'simplify' an expression like $5n^2 + 7n + 1$, and you may need to emphasise that only like terms can be collected together.

C Bracket pairs (p 136)

◊ The area diagram may remind pupils of the similar method for multiplying two-digit numbers:

$$\begin{array}{c|cc} \times & 30 & 2 \\ \hline 30 & 900 & 60 \\ 8 & 240 & 16 \end{array}$$

$$32 \times 38 = 900 + 60 + 240 + 16 = 1216$$

◊ Both the area diagram and the table can be linked to the distributive law. First think of one of the brackets as if it were a single symbol:

$$(n + 2) \, \boxed{(n + 8)} = n \, \boxed{(n + 8)} + 2 \, \boxed{(n + 8)}$$

◊ Discuss the difference between the statement $(n + 1)(n + 3) = n^2 + 4n + 3$ and a statement such as $n^2 = 9$. Pupils can substitute various values for n in the first statement and help confirm that it is true for all values of n.

D More brackets (p 137)

◊ Pupils can be asked to try multiplying out the brackets from the products A, B and C. Their solutions and methods can be compared.

An area diagram may be confusing when one or both brackets involve subtraction. One approach is to replace, for example, $(x - 4)$ by $(x + {}^{-}4)$. Another is to use the distributive law: $(x - 4)(x + 5) = x(x + 5) - 4(x + 5)$

E Number patterns and proof (p 138)

◊ Emphasise that producing many lines of a pattern all with the result 2 is not a proof that the result will always be 2. The pattern may break down for a value not yet tried.

A good example where this happens is the expression $n^2 + n + 41$. This expression is prime for all values of n up to $n = 39$, but not for $n = 40$.

◊ The expression for the nth line of the first pattern is
$$(n + 1)(n + 2) - n(n + 3)$$
You may wish to point out that n could be chosen to stand for any of the four numbers in each expression. However, other choices complicate the manipulation. For example, if n stands for the first number then the general expression becomes $n(n + 1) - (n - 1)(n + 2)$.

◊ The expression for the nth line of the second pattern is
$$(n + 1)(n + 5) - n(n + 6)$$
which simplifies to $n + 10$, showing that the result is always 10 more than the third number.

F Opposite corners (p 139)

F1, 2 By choosing n to stand for the number in the top left-hand corner, pupils can prove that the 'opposite corners number' will always be 40.

Encourage high attainers, when generalising to different sized squares on the ten-column grid, to look first for a general rule that links the 'opposite corners number' with the size of the square. Then in F2 they can think about the general rule that links the 'opposite corners number' with the size of the square and the number of columns.

G True, iffy, false? (p 139)

◊ Emphasise that to prove a statement is **sometimes** true (iffy), you only need to find one value for n that makes the statement true and one that makes the statement false. However, to prove that a statement is **always** true or **never** true a more general argument is needed.

G1 In part (b) the fact that $(n + 1)(n + 2)$ is always even can be proved in either of two ways:
- $(n + 1)$ and $(n + 2)$ are consecutive numbers, so one must be odd and the other even, so their product must be even.
- $(n + 1)(n + 2) = n^2 + 3n + 2$.
  If n is odd, this expression is odd + odd + even = even.
  If n is even, it is even + even + even = even.

Some statements ((f) and (i)) involve factorising.

Statement (j) is difficult to prove never true. Ask pupils to consider the expression $n^2 + 10n + 25$ and to try to prove that it is always a square number.

H Mixed questions (p 139)

A Simplifying: review (p 134)

A1 (a) $8p + 3$ (b) $11x - 6$ (c) $17 - 3n$

A2 (a) $3p - 2$ (b) $15 - m$ (c) $4h + 1$
 (d) $13 - 2h$ (e) $6x - 1$ (f) $3 + 2y$

A3 (a) $\mathbf{6}y - (3 - 2y) = 8y - 3$
 (b) $\mathbf{6} - (2p - 1) = 7 - 2p$

A4 (a) $^-3x - 12$ (b) $^-8 + 6a$
 (c) $3h - 7$ (d) $25 - 5b$
 (e) $^-12$ (f) $20 - 15d$

A5 (a) $10 - 3(2x + \mathbf{1}) = 7 - 6x$
 (b) $6k - 3(1 + \mathbf{2}k) = {}^-3$
 (c) $\mathbf{14}y - 2(5 - 3y) = 20y - 10$
 (d) $\mathbf{26} - 3(2p + 1) = 23 - 6p$

B Further simplification (p 135)

B1 (a) $4x^2 + 5x$ (b) $4h^2 + h + 5$
 (c) $6m^2 - 3m + 9$ (d) $10 - n^2$
 (e) $2b^2 - 4b$ (f) $k^2 - 6k + 5$

B2 (a) $n^2 - n$ (b) $2m^2 + 15m$
 (c) $40 + 5h - h^2$ (d) $4b^2 - b$
 (e) $17x - x^2$ (f) $4g^2$

B3 (a) $2n^2 + 4n$ (b) $b^2 + 4b - 5$
 (c) $3m^2 - 3m$ (d) $5y^2 - 18y$

B4 (a) $d^2 - d$ (b) $3n$
 (c) $2x^2 - 2x$ (d) $p^2 - 8p + 15$

B5 (a) $3x^2 + 14x$ (b) $11y^2 - 2y$

B6 (a) $y(y + 3) + \mathbf{5}(y + 3) = y^2 + 8y + \mathbf{15}$
 (b) $\mathbf{d}(d - 1) + \mathbf{5}(d - 1) = d^2 + 4d - \mathbf{5}$
  (or $\mathbf{5}(d - 1) + \mathbf{d}(d - 1)$)
 (c) $\mathbf{2x}(x + 1) - \mathbf{x}(x + 5) = x^2 - 3x$

C Bracket pairs (p 136)

Each product and sum can be written in a different order.
For example $(n + 3)(n + 1)$ is the same as $(n + 1)(3 + n)$.

C1 (a) $(n + 2)(n + 5)$, $n^2 + 7n + 10$
(b) $(n + 3)(n + 3)$, $n^2 + 6n + 9$
(c) $(n + 1)(n + 10)$, $n^2 + 11n + 10$

C2 (a) $x^2 + 8x + 12$ (b) $x^2 + 13x + 12$
(c) $x^2 + 5x + 4$ (d) $x^2 + 4x + 4$
(e) $x^2 + 9x + 18$ (f) $x^2 + 11x + 18$

C3 (a) $(x + 5)^2 = (x + 5)(x + 5) =$
$x^2 + 10x + 25$
(b) $x^2 + 14x + 49$

C4 (a) The pupil's explanation, for example, 'The pupil has just squared the 'x' and the '4'.'
(b) $x^2 + 8x + 16$

C5 (a) $(n + 3)(n + 4)$ (b) $(n + 1)(n + 5)$
(c) $(n + 1)(n + 6)$ (d) $(n + 2)(n + 3)$

C6 (a) $a + 7$ (b) $b + 4$

C7 (a) $p + 10$
(b) This rectangle is also a square.

C8 (a) $(n + 5)(\boldsymbol{n + 3}) = n^2 + 8n + 15$
(b) $(\boldsymbol{n + 7})(n + 3) = n^2 + 10n + 21$
(c) $(k + 2)(\boldsymbol{k + 11}) = k^2 + \boldsymbol{13}k + 22$
(d) $(\boldsymbol{k + 2})(k + 4) = k^2 + 6k + \boldsymbol{8}$

C9 (a) $(\boldsymbol{a + 2})(\boldsymbol{a + 1}) = a^2 + 3a + 2$
(b) $(\boldsymbol{b + 8})(\boldsymbol{b + 3}) = b^2 + 11b + 24$
(c) $(\boldsymbol{c + 1})(\boldsymbol{c + 1}) = c^2 + 2c + 1$
(d) $(\boldsymbol{d + 13})(d + 2) = d^2 + 15d + 26$

C10 (a) It is not possible to complete the expression. The pupil's explanation, for example, 'You cannot find two integers whose sum is 2 and whose product is 5.'

(b) It is possible when two integers exist so that their sum is a and their product is b.

D More brackets (p 137)

D1 (a) $x^2 + 3x - 10$ (b) $x^2 + 8x - 9$
(c) $x^2 - 3x - 18$ (d) $x^2 - 49$
(e) $x^2 - 7x + 12$ (f) $x^2 - 6x + 5$
(g) $x^2 - 1$ (h) $x^2 - 4x + 4$
(i) $x^2 - 6x + 9$

D2 (a) $(n + 3)(n - 1)$ (b) $(n + 1)(n - 2)$
(c) $(n - 2)(n - 3)$ (d) $(n + 3)(n - 3)$

D3 (a) $(n + 5)(\boldsymbol{n - 3}) = n^2 + 2n - 15$
(b) $(\boldsymbol{m - 3})(m - 6) = m^2 - \boldsymbol{9}m + 18$
(c) $(\boldsymbol{a - 2})(\boldsymbol{a - 5}) = a^2 - 7a + 10$
(d) $(\boldsymbol{b - 6})(b + 2) = b^2 - 4b - 12$
(e) $(\boldsymbol{c - 8})(\boldsymbol{c + 1}) = c^2 - 7c - 8$
(f) $(\boldsymbol{d - 5})^2 = d^2 - 10d + 25$

D4 A, D

E Number patterns and proof (p 138)

E1 A (a) $4 \times 4 - 1 \times 7 = \boldsymbol{9}$
$5 \times 5 - 2 \times 8 = \boldsymbol{9}$
$6 \times 6 - 3 \times 9 = \boldsymbol{9}$
$\boldsymbol{7} \times \boldsymbol{7} - 4 \times \boldsymbol{10} = \boldsymbol{9}$

(b) $(\boldsymbol{n + 3}) \times (\boldsymbol{n + 3}) - n \times (\boldsymbol{n + 6})$
$= (n^2 + 6n + 9) - (n^2 + 6n) = 9$

(c) It proves the result for each line will be 9.

B (a) $4 \times 6 - 1 \times 8 = \boldsymbol{16}$
$5 \times 7 - 2 \times 9 = \boldsymbol{17}$
$6 \times 8 - 3 \times 10 = \boldsymbol{18}$
$\boldsymbol{7} \times \boldsymbol{9} - 4 \times \boldsymbol{11} = \boldsymbol{19}$

(b) $(\boldsymbol{n + 3}) \times (\boldsymbol{n + 5}) - n \times (\boldsymbol{n + 7})$
$= (n^2 + 8n + 15) - (n^2 + 7n)$
$= n + 15$

(c) It proves the result for the nth line is 15 more than n.

E2 (a) $6 \times 3 - 8 \times 1 = \mathbf{10}$
$7 \times 4 - 9 \times 2 = \mathbf{10}$
$8 \times 5 - 10 \times 3 = \mathbf{10}$

(b) The nth line of this pattern is
$(n + 5) \times (n + 2) - (n + 7) \times n$
which simplifies to
$(n^2 + 7n + 10) - (n^2 + 7n) = 10$

(c) This proves that the result for each line will always be 10.

E3 The pupil's pattern with result 6

E4 (a) $7 \times 5 - 1 \times 9 = \mathbf{26}$
$8 \times 6 - 2 \times 10 = \mathbf{28}$
$9 \times 7 - 3 \times 11 = \mathbf{30}$

(b) The nth line of this pattern is
$(n + 6)(n + 4) - n(n + 8)$
which simplifies to
$(n^2 + 10n + 24) - (n^2 + 8n)$
$= 2n + 24$

(c) The pupil's explanation, for example:
$2n$ and 24 are both even for all values of n and the sum of two even numbers is even so the result for each line is even.

***E5** (a) $2 \times 2 - 1 \times 1 = 3$
$3 \times 3 - 2 \times 2 = \mathbf{5}$
$4 \times 4 - 3 \times 3 = \mathbf{7}$
$\mathbf{5 \times 5 - 4 \times 4 = 9}$

(b) The pupil's proof. A possible proof is:
The pattern above gives the differences between pairs of consecutive square numbers.
The nth line of this pattern is
$(n + 1)(n + 1) - n^2 =$
$(n^2 + 2n + 1) - n^2 = 2n + 1$.
$2n + 1$ is one more than an even number ($2n$) and so must be odd.

***E6** The pupil's proof. A possible proof is:
The sum of two consecutive square numbers can be written
$n^2 + (n + 1)(n + 1)$
which simplifies to $2n^2 + 2n + 1$.
$2n^2$ and $2n$ are even so $2n^2 + 2n$ is even (being the sum of two even numbers).

Hence $2n^2 + 2n + 1$ must be odd (being one more than an even number).

F Opposite corners (p 139)

F1 (a) The result is always 40.

(b) The pupil's proof (see guidance notes for a possible proof)

(c) The pupil's investigation

F2 The pupil's investigation

G True, iffy, false? (p 139)

G1 (a) Sometimes   (b) Always
(c) Never   (d) Never
(e) Always   (f) Always
(g) Sometimes   (h) Sometimes
(i) Never   (j) Never

H Mixed questions (p 139)

H1 (a) $a + b, a - b$

(b) Area of rectangle $= (a + b)(a - b)$
Coloured area in first diagram $=$
$a^2 - b^2$
So $a^2 - b^2 = (a + b)(a - b)$

(c) $(a + b)(a - b) = a^2 - ab + ba - b^2$

H2 (a) $x^2 + 7x + 12 = x^2 + 40$

(b) $7x + 12 = 40$
           $x = 4$

H3 (a) $x = 4$ (b) $x = 28$
(c) $x = {}^-10$ (d) $x = {}^-4$

H4 $r = 5.6$

What progress have you made? (p 140)

1 (a) $n^2 + 2n - 8$ (b) $p^2 + 11p$

2 (a) $n^2 + 9n + 20$ (b) $x^2 + 9x + 8$

3 $a + 2$

4 (a) $n^2 + 2n - 24$ (b) $x^2 - 11x + 18$

5 nth line, left-hand side

$$= (n + 1)(n + 5) - n(n + 6)$$
$$= n^2 + 6n + 5 - n^2 - 6n$$
$$= 5$$

Practice booklet

Sections A and B (p 57)

1 (a) $4p - 7$ (b) $11 - 2k$

   (c) $^-2 - n$ (d) $7y - 5$

2 (a) $7p - 6$ (b) $5 - 3h$

   (c) $6 - 7k$ (d) $n + 14$

3 (a) $p^2 - p$ (b) $h^2 - 12h + 4$

   (c) $k^2 - 3k$ (d) $11n - n^2$

4 (a) $3a^2 + 10a$ (b) $2n^2 + 11n$

5 (a) $p^2 + 4p + 6$ (b) $a^2 + 2a - 6$

   (c) $3a^2 + 4a$ (d) $2x^2 - 8$

6 (a) $a(a + 2) + \mathbf{3}(a + 2) = a^2 + 5a + \mathbf{6}$

   (b) $\mathbf{2}(p - 2) + p(p + 4) = p^2 + 6p - \mathbf{4}$

Sections C and D (p 58)

1 (a) $x^2 + 6x + 8$ (b) $x^2 + 9x + 8$

   (c) $x^2 + 22x + 121$ (d) $x^2 - 3x - 18$

   (e) $x^2 + 7x - 18$ (f) $x^2 - 7x - 18$

   (g) $x^2 - 9x + 20$ (h) $x^2 - 10x + 21$

   (i) $x^2 - 6x + 9$

2 (a) $(a + 3)(a + 2)$ (b) $(a - 3)(a - 2)$

   (c) $(a + 3)(a - 2)$ (d) $(a - 3)(a + 2)$

   (e) $(a - 3)(a - 4)$ (f) $(a - 6)(a - 2)$

   (g) $(a + 3)(a - 3)$ (h) $(a + 4)(a - 4)$

Section E (p 58)

1 (a) $5 \times 6 - 1 \times 10 = \mathbf{20}$

     $6 \times 7 - 2 \times 11 = \mathbf{20}$

     $7 \times 8 - 3 \times 12 = \mathbf{20}$

   (b) $(n + 4)(n + 5) - n(n + 9) =$
     $(n^2 + 9n + 20) - (n^2 + 9n) = 20$
     The result will always be 20.

2 The pupil's pattern, for example,
   $(n + 1)(n + 4) - n(n + 5)$

3 (a) **A** $(n + 1)(n + 3) - n(n + 2) =$
           $(n^2 + 4n + 3) - (n^2 + 2n) =$
           $2n + 3$

       B $(n + 2)(n + 3) - n(n + 3) =$
           $(n^2 + 5n + 6) - (n^2 + 3n) =$
           $2n + 6$

       C $(n + 1)(n + 5) - n(n + 6) =$
           $(n^2 + 6n + 5) - (n^2 + 6n) = 5$

   (b) (i) C   (ii) B   (iii) A (and C)

Section F (p 59)

1 (a)

   (b) $(n + 9)(n + 11) - n(n + 20) =$
     $(n^2 + 20n + 99) - (n^2 + 20n) = 99$
     The diamond number will always
     be 99.

2 The number would be 63.

   $(n + 7)(n + 9) - n(n + 16) =$
   $(n^2 + 16n + 63) - (n^2 + 16n) = 63$

㉑ Quadratic sequences

This unit leads to finding the formula for the nth term of a quadratic sequence. There is plenty of scope for using a spreadsheet here, though pupils should be able to find the nth term without using calculating aids.

The techniques in this unit may be found very useful if a pupil is trying to find an algebraic summary of an open-ended piece of work.

	Optional Spreadsheet program
Practice booklet pages 60 and 61	

Ⓐ Sequences (p 141)

This section gives an opportunity to revise the basic ideas of sequences, including expressions for the nth term.

Ⓑ Linear sequences (p 142)

This section revises finding the nth term of a linear sequence by the difference method.

Ⓒ Second differences (p 143)

The example may be used to introduce second differences. The fact that a quadratic sequence (one whose nth term is a quadratic function) has constant second differences will come out in the next section.

Ⓓ Investigating quadratic sequences (p 144)

This investigation may be done in pairs or small groups, perhaps with input from the teacher at key points.

Ⓔ Finding the formula for a quadratic sequence (p 145)

Optional: spreadsheet program

◊ At the start of this section you way wish to bring together and discuss the results of the previous section. Question E1 could be used to go through the method.

◊ For higher attainers you could give a more rigorous algebraic explanation of the reason why the second differences are constant. It is a good example of multiplying out brackets.

If the nth term is $an^2 + bn + c$,
the $(n + 1)$th term is $a(n + 1)^2 + b(n + 1) + c$,
the first difference, between the nth and $(n + 1)$th terms, is $2an + a + b$,
the next value of the first difference is $2a(n + 1) + a + b$,
giving a second difference of $2a$.

◊ For your benefit, here is a quick way to find the values of a, b and c:

Extend the sequence backwards to give a '0th term', and extend the first differences backwards as well.

The value of the 0th term is obviously c.

The new first value of the first differences is $a + b$ (see proof above), so having found a from the second differences it is easy to work out b.

Example: 10, 17, 28, 43, …

Extend backwards:	**7**	10	17	28	43	…
First differences:		**3**	7	11	15	…
Second differences			**4**	4	4	…

So $c = 7$, $a = 2$ and $a + b = 3$, from which $b = 1$.
The formula for the nth term is $2n^2 + n + 7$.

Using a spreadsheet

◊ Some spreadsheets (including later versions of Excel) allow you to head a column with a letter, n in this case, and then express spreadsheet formulas in terms of n.

Here is an example of how a spreadsheet might be used to reinforce the work on quadratic sequences. It is an activity for two people.

• The first column is used for values of n, i.e. 1, 2, 3, 4, …

• One person devises a quadratic sequence (away from the spreadsheet) and then enters the terms in the second column, e.g. 9, 13, 19, 27, …

• The second person enters and fills down formulas for first differences and second differences.

	A	B	C	D	E
1	n	sequence	1st diffs	2nd diffs	
2	1	9			
3	2	13	4		
4	3	19	6	2	
5	4	27	8	2	
6					

- The second person then works out the formula for the nth term (away from the spreadsheet).
- The formula is tested by entering it in the next column and comparing the result with the given sequence.

	A	B	C	D	E
1	n	sequence	1st diffs	2nd diffs	n^2 + n + 7
2	1	9			9
3	2	13	4		13
4	3	19	6	2	19
5	4	27	8	2	27
6					

A Sequences (p 141)

A1 (a) 19, 23 (add 4 to previous term)

(b) 34, 41 (add 7 to previous term)

(c) 19, 25 (the difference increases by 1)

(d) 9, 4 (subtract 5 from previous term)

(e) 26, 37 (the difference increases by 2)

(f) 96, 192 (double the previous term)

(g) 63, 127 (double the previous term and add 1, or double the difference each time)

(h) 78, 105 (the difference increases by 4)

A2 (a)

n	1	2	3	4	5	6	7	8
$3n + 1$	4	**7**	10	**13**	16	**19**	**22**	**25**

(b) The 25th term

A3 (a) 1, 7, 13, 19, 25

(b) $n = 17$

A4 (a) 0, 3, 8, 15, 24

(b) $n = 11$

A5 (a) 60, 30, 20, 15, 12

(b) (i) $n = 12$ (ii) $n = 120$

A6 (a) $0, \frac{1}{3}, \frac{1}{2}, \frac{3}{5}, \frac{2}{3}$

(b) $n = 7$

B Linear sequences (p 142)

B1 (a)

n	1	2	3	4	5	6	7	8
$4n + 3$	7	11	15	**19**	**23**	**27**	**31**	**35**
Differences		4	4	**4**	**4**	**4**	**4**	**4**

(b)

n	1	2	3	4	5	6	7	8
$5n - 1$	4	9	14	19	24	29	34	39
Differences		5	5	5	5	5	5	5

(c)

n	1	2	3	4	5	6	7	8
$7n + 4$	11	18	25	32	39	46	53	60
Differences		7	7	7	7	7	7	7

(d) (i) 2   (ii) 3   (iii) 8   (iv) 2   (v) ⁻4

B2 (a)

n	1	2	3	4	5	6	7	8
Sequence	8	14	20	26	32	38	44	50
Differences		6	**6**	**6**	**6**	**6**	**6**	**6**

(b) $6n + 2$

(c) For example, when $n = 5$, $6n + 2 = 32$

B3 (a) $4n + 1$ (b) $6n - 3$

(c) $0.5n + 4.5$ (d) ⁻$3n + 7$ or $7 - 3n$

(e) ⁻$4n + 3$ or $3 - 4n$

(f) Non-linear, $n^2 + 1$

C Second differences (p 143)

C1 (a)

Sequence	4		6		9		13		18		**24**		**31**
1st diff		2		3		4		5		6		7	
2nd diff			1		1		1		1		1		

(b)

Sequence	4		10		18		28		40		54		**70**		**88**
1st diff		6		8		**10**		**12**		**14**		**16**		**18**	
2nd diff			2		2		2		2		2		2		

(c)

Sequence	0		7		18		33		52		**75**		**102**
1st diff		**7**		**11**		**15**		**19**		**23**		27	
2nd diff			**4**		**4**		**4**		**4**		**4**		

(d)

Sequence	3.4		4.7		6.2		7.9		**9.8**		**11.9**
1st diff		**1.3**		**1.5**		**1.7**		**1.9**		2.1	
2nd diff			**0.2**		**0.2**		**0.2**		**0.2**		

C2 (a)

n	1		2		3		4		5		6
Sequence	**2**		**6**		**12**		20		**30**		**42**
1st diff		4		6		8		**10**		**12**	
2nd diff			**2**		**2**		**2**		**2**		

(b)

n	1		2		3		4		5		6
Sequence	**9**		**12**		**19**		30		**45**		**64**
1st diff		3		7		**11**		**15**		19	
2nd diff			**4**		**4**		**4**		**4**		

(c)

n	1		2		3		4		5		6
Sequence	**15**		**20**		**24**		27		**29**		**30**
1st diff		**5**		**4**		3		**2**		1	
2nd diff			**⁻1**		**⁻1**		**⁻1**		**⁻1**		

(d)

n	1		2		3		4		5		6
Sequence	**⁻5**		**7**		**18½**		**29½**		**40**		50
1st diff		12		**11½**		**11**		**10½**		10	
2nd diff			**⁻½**		**⁻½**		**⁻½**		**⁻½**		

C3 (a)

Sequence	3		10		29		66		127		218		**345**		**514**
1st diff		7		19		37		61		91		127		169	
2nd diff			12		18		24		30		36		42		
3rd diff				6		6		6		6		6			

(b)

Sequence	0		1		4		10		20		35		**56**		**84**
1st diff		1		3		6		10		15		21		28	
2nd diff			2		3		4		5		6		7		
3rd diff				1		1		1		1		1			

D Investigating quadratic sequences (p 144)

The completed tables are as follows.

n	1		2		3		4		5		6		7
n^2	1		4		9		16		25		36		49
1st diff		3		5		**7**		**9**		**11**		**13**	
2nd diff			**2**		**2**		**2**		**2**		**2**		

n	1		2		3		4		5		6		7
$3n^2$	3		12		27		48		**75**		**108**		**147**
1st diff		9		15		21		**27**		**33**		**39**	
2nd diff			**6**		**6**		**6**		**6**		**6**		

Second difference is 2 × coefficient of n^2

n	1		2		3		4		5		6		7
$3n^2 + 7$	10		19		34		55		**82**		**115**		**154**
1st diff		9		15		21		**27**		**33**		**39**	
2nd diff			**6**		**6**		**6**		**6**		**6**		

Constants and terms in n have no effect on the second differences.

E Finding the formula for a quadratic sequence (p 145)

E1 (a)

n	1		2		3		4		5		6		7
Sequence	4		15		32		55		84		119		160
1st diff		11		17		**23**		**29**		**35**		**41**	
2nd diff			**6**		**6**		**6**		**6**		**6**		

(b)

n	1		2		3		4		5		6		7
Sequence	4		15		32		55		84		119		160
Seq – $3n^2$	1		3		5		7		9		11		13
1st diff		**2**		**2**		**2**		**2**		**2**		**2**	

(c) The nth term of 'sequence – $3n^2$' is $2n - 1$.

(d) The nth term is $3n^2 + 2n - 1$

E2 (a)

n	1		2		3		4		5		6		7
Sequence	7		17		31		49		71		97		127
1st diff		10		14		**18**		**22**		**26**		**30**	
2nd diff			**4**		**4**		**4**		**4**		**4**		

(b) $a = 2$

(c)

n	1		2		3		4		5		6		7
Sequence	7		17		31		49		71		97		127
Seq – $2n^2$	**5**		**9**		**13**		**17**		**21**		**25**		**29**
1st diff		**4**		**4**		**4**		**4**		**4**		**4**	

(d) Sequence – $2n^2$ has nth term $4n + 1$
Original has nth term $2n^2 + 4n + 1$

(e) 8th term = 161, 9th term = 199

E3 The nth term of each sequence is

(a) $4n^2 + 3n$ (b) $2n^2 - n + 5$

(c) $4n^2 + 5n - 9$

(d) $2n^2 - 3n + 1$ (e) $3n^2 + n - 7$

(f) $\frac{3}{4}n^2 + \frac{1}{4}$

*E4 (a) n^3 (b) $n^3 + 2$

(c) 2^n (d) 10^n

(e) $10^n + n$ (f) $10^n + n^2$

(g) $\frac{1}{n}$ (h) $\frac{1}{n+1}$

(i) $\frac{n}{n+1}$ (j) $\frac{2n-1}{3n-2}$

What progress have you made? (p 146)

1

Sequence	2		2		4		8		14		**22**		**32**
1st diff		**0**		**2**		**4**		**6**		**8**		**10**	
2nd diff			**2**		**2**		**2**		**2**		**2**		

2 The first five terms are 0, 5, 16, 33, 56.

3

n	1		2		3		4		5		6
Sequence	8		26		54		92		140		198
1st diff		**18**		**28**		**38**		**48**		**58**	
2nd diff			**10**		**10**		**10**		**10**		

The coefficient of n^2 is therefore 5.

Subtracting $5n^2$ from each term, we get

Sequence	8	26	54	92	140	198
$5n^2$	**5**	**20**	**45**	**80**	**125**	**180**
Seq $- 5n^2$	**3**	**6**	**9**	**12**	**15**	**18**

By inspection, taking off $5n^2$ leaves $3n$, so the nth term of the sequence is $5n^2 + 3n$.

Practice booklet

Sections B and C (p 60)

1 (a) 4, 9, 14, **19**, **24** $5n - 1$

(b) 5, 8, **11**, **14**, **17** $3n + 2$

(c) **12**, **8**, 4, 0, $^-4$ $^-4n + 16$

(d) **4.5**, 5, **5.5**, 6, **6.5**, 7 $0.5n + 4$

(e) **12.6**, **11.3**, 10, **8.7**, 7.4 $^-1.3n + 13.9$

(f) 7.5, **9**, **10.5**, 12, **13.5** $1.5n + 6$

(g) 2.1, 2.8, **3.5**, **4.2**, **4.9** $0.7n + 1.4$

(h) $5\frac{2}{3}$, 5, $4\frac{1}{3}$, $3\frac{2}{3}$, 3 $-\frac{2}{3}n + 6\frac{1}{3}$

2 (a)

n	1		2		3		4		5		6
Sequence	1		**4**		**9**		**16**		**25**		**36**
1st diff		**3**		**5**		**7**		**9**		**11**	
2nd diff			**2**		**2**		**2**		**2**		

(b)

n	1		2		3		4		5		6
Sequence	5		**3**		**5**		**11**		**21**		**35**
1st diff		**$^-2$**		**2**		**6**		**10**		**14**	
2nd diff			**4**		**4**		**4**		**4**		

3 (a) 1, 2, 4, 9, 19, 36, **62**, **99**

(b) 1, 3, 7, 15, 31, **63**, **127**
if it is $2^n - 1$ or
1, 3, 7, 15, 31, **61**, **113**
if the 4th differences are constant

Section E (p 61)

1 (a)

4		7		12		19		28
	3		5		7		9	
		4		4		4		

(b)

5		8		15		26		41
	3		7		11		15	
		4		4		4		

(c)

$^-1$		10		27		50		79
	11		17		23		29	
		6		6		6		

(d)

$2\frac{1}{2}$		6		$10\frac{1}{2}$		16		$22\frac{1}{2}$
	$3\frac{1}{2}$		$4\frac{1}{2}$	$5\frac{1}{2}$		$6\frac{1}{2}$	
		1		1		1		

(e)

$-\frac{3}{4}$		2		$5\frac{1}{4}$		9		$13\frac{1}{4}$
	$2\frac{3}{4}$		$3\frac{1}{4}$	$3\frac{3}{4}$		$4\frac{1}{4}$	
		$\frac{1}{2}$		$\frac{1}{2}$		$\frac{1}{2}$		

(f)

$^-1$		0		0		$^-1$		$^-3$
	1		0		$^-1$		$^-2$	
		$^-1$		$^-1$		$^-1$		

2 (a) 2   (b) 4   (c) 1   (d) 10   (e) $^-6$ (f) $\frac{1}{2}$

3 (a)

n	1		2		3		4		5		6
Sequence	$^-1$		0		3		8		**15**		**24**
1st diff		1		**3**		**5**		**7**		**9**	
2nd diff			2		**2**		**2**		**2**		

The nth term is $n^2 - 2n$.

(b)

n	1		2		3		4		5		6
Sequence	5		10		19		**32**		**49**		**70**
1st diff		**5**		**9**		**13**		**17**		**21**	
2nd diff			**4**		**4**		**4**		**4**		

The nth term is $2n^2 - n + 4$.

4 (a)

6		9		14		**21**		**30**		41
	3		5		7		9		**11**	
		2		2		2		2		

The nth term is $n^2 + 5$.

(b)

5		12		21		32		**45**		60
	7		9		11		13		15	
		2		2		2		2		

The nth term is $n^2 + 4n$.

(c)

1		7		17		31		49		**71**
	6		10		14		18		22	
		4		4		4		4		

The nth term is $2n^2 - 1$.

(d)

5		14		27		44		**65**		90
	9		13		17		21		25	
		4		4		4		4		

The nth term is $2n^2 + 3n$.

Review 3 (p 147)

1 (a) $x = \dfrac{y-5}{3}$ (b) $x = 3y + 4$

(c) $x = 5(y - 7)$

2 (a) 4.5 (b) 13.5 cm

3 (a) (i) $x = \pm 3.2$, approximately

(ii) $x = \pm 2.5$, approximately

(b) $y = \frac{1}{2}x^2 + 1$

4 (a) $x \geq {}^-3$ (b) $x < 3$

(c) $^-2 \leq x \leq 4$ (d) $^-1 < x \leq 2$

5 (a) $13x - 27$ (b) $x^2 + 11x + 28$

(c) $x^2 + 6x - 27$

6 (a) n^2

(b) (i) $n^2 + 30$ (ii) $\frac{1}{2}n^2$

(iii) $\frac{1}{2}n^2 + \frac{1}{2}$ (iv) $(n-1)^2$

7 (a) Perimeter of each semicircular
end $= \frac{1}{2}\pi s$

P = total perimeter $= \pi s + 2s$
$= (\pi + 2)s$

(b) $s = \dfrac{P}{\pi + 2}$

(c) Radius of each end $= \frac{1}{2}s$
Total area of two ends $= \pi(\frac{1}{2}s)^2$
$= \dfrac{\pi s^2}{4}$

Area of square $= s^2$
Total area $= \frac{\pi}{4}s^2 + s^2 = (\frac{\pi}{4} + 1)s^2$

(d) $s = 5.0$

8 (a) 0.4 (b) 7.8 cm

9 (a) $^-2, ^-1, 0, 1, 2, 3$

(b) $^-2, ^-1, 0, 1$

(c) $^-3, ^-2, ^-1, 0, 1, 2, 3$

10 (a) $8 \times 9 - 5 \times 12 = 12$
$9 \times 10 - 6 \times 13 = 12$

(b) $(n + 3)(n + 4) - n(n + 7) = \ldots$

(c) $(n + 3)(n + 4) - n(n + 7)$
$= n^2 + 7n + 12 - n^2 - 7n$
$= 12$

11 (a) 1.82 (b) 2.96 (c) 22.1

12 18.7 cm

13 $V = 5 - \frac{1}{5}I$ or $V = {}^-0.2I + 5$, etc.

14 (a) 40 (b) 39 (c) 30

(d) No change (e) 0.25

15 (a)

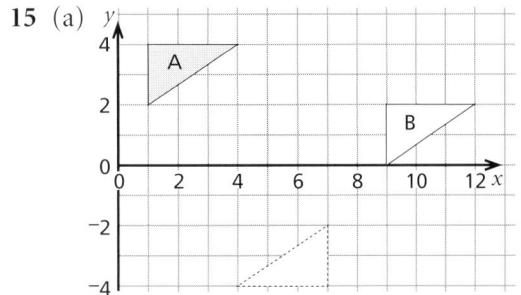

(b) Translation $\begin{bmatrix} 8 \\ -2 \end{bmatrix}$

(c)

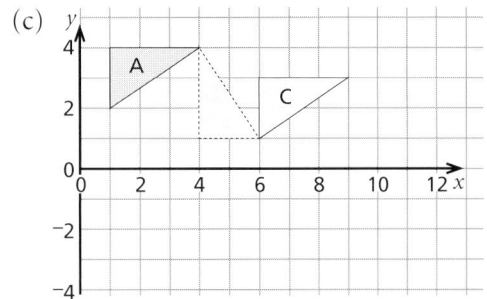

(d) Translation $\begin{bmatrix} -5 \\ 1 \end{bmatrix}$

16 (a) 900° (b) 157.5°

17 (a) $x = {}^-3$ (b) $x = {}^-5$

(c) $x = 1\frac{1}{2}$ (d) $x = 4\frac{1}{2}$

18 (a) $\frac{7}{15}$ (b) $\frac{1}{10}$ (c) $\frac{13}{30}$

19 $b = \dfrac{P}{2} - a$

*20 $x = 30$ cm

*21 1, 1, 1, 2, $\boxed{2, \ 3,}$ $\boxed{4,}$ $\boxed{5}$

Each term is the sum of the terms three
before it and two before it.

Explanation: Suppose we want to make, say, 17p. We can use all the ways to make 14p and add a 3p at the end, or all the ways to make 15p and add a 2p.

Mixed questions 3 (Practice booklet p 62)

1 (a) $A = 35$ (b) $h = 6$

   (c) $a = \dfrac{2A}{h} - b$ or $a = \dfrac{2A - hb}{h}$

2 (a) 2.5   (b) 12 cm

   (c) 35 cm

3 (a)

x	-4	-3	-2	-1	0	1	2	3	4
y	11	4	-1	-4	-5	-4	-1	4	11

   (a), (c)

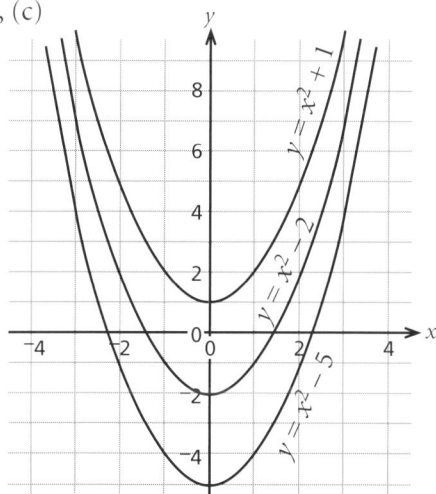

   (b) $x = \pm\,2.2$

4 (a) -1, 0, 1, 2, 3   (b) -2, -1, 0, 1, 2, 3
   (c) 8, 9, 10, -8, -9, -10

5 (a) $2a + 2$ (b) $16b - 15$
   (c) $-c - 2c^2$ (d) $d^2 + 10d + 21$
   (e) $e^2 - e - 12$ (f) $f^2 - 16f + 64$

6 (a) $n^2 - 1$ (b) $2n^2 - 3n + 7$

7 (a) $150\pi\,\text{cm}^2$ (b) $250\pi\,\text{cm}^3$

8 13.4 km/h

9 The pupil's accurate diagram of rectangle ABCD and that part of the perpendicular bisector of AC that lies inside the rectangle.

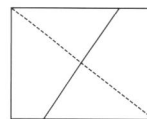

10 (a) (i) £122.70
       (ii) $177.91
   (b) €562.07

11 24°

12 (a) (i) $x = \frac{2}{3}$, AB = AC = 7, BC = 5
       (ii) $x = 1$, AB = BC = 6, AC = 8
   (b) There is no solution to the equation

13 20.5 cm² to 1 d.p.

14 45

*15 (a)

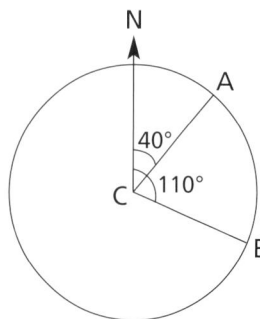

   (b) 165°

*16 $\frac{3}{8}$